SpringerBriefs in Physics

For further volumes:
http://www.springer.com/series/8902

Zeev Zalevsky

Editor

Super-Resolved Imaging

Geometrical and Diffraction Approaches

 Springer

Editor
Zeev Zalevsky
School of Engineering
Bar-Ilan University
Ramat-Gan 52900, Israel
zalevsz@macs.biu.ac.il

ISSN 2191-5423   e-ISSN 2191-5431
ISBN 978-1-4614-0832-1   e-ISBN 978-1-4614-0833-8
DOI 10.1007/978-1-4614-0833-8
Springer New York Dordrecht Heidelberg London

Library of Congress Control Number: 2011934035

Printed on acid-free paper

Springer is part of Springer Science+Business Media (www.springer.com)

Preface

Super resolution is one of the most fascinating and applicable fields in optical data processing. The urge to obtain highly resolved images using low-quality imaging optics and detectors is very appealing.

The field of super resolution may be categorized into two groups: diffractive and geometrical super resolution. The first deals with overcoming the resolution limits that are dictated by diffraction laws and related to the numerical aperture of the imaging lens. The second deals with overcoming the limitation determined by the geometrical structure of the detector array.

Various techniques have been developed to deal with both types of resolution improvements. In all approaches, the spatial resolution improvement needs the object to exhibit some sort of constraint (such as monochromaticity, slow variation with time, single polarization, etc.), related with an unused dimension of the object. The improvement is thus made at the price of sacrificing unused degrees of freedom in the other domains as time, wavelength, polarization, or field of view.

The methods pursuing super resolution utilize masks having diffractive features. They are classified here according to the nature of their structure:

1. Possessing full/piecewise periodicity
2. Spatially finite repeating random structures/random structure with finite period
3. Random structure with infinite period

The book is thus organized in the following way. Chapter 1 briefly presents the relevant theoretical background. Chapter 2 discusses several super resolution methods implementing diffractive masks having a certain degree of periodicity. In Chapter 3, we explore techniques utilizing diffractive masks having structures with a finite random period. Finally, in Chapter 4, the mask becomes fully random.

Ramat-Gan, Israel   Zeev Zalevsky

Contents

Contributors

Amikam Borkowski School of Engineering, Bar-Ilan University, Ramat-Gan, Israel

Javier García Departamento de Óptica, Universitat de Valencia, Burjassot, Spain

Bahram Javidi Department of Electrical and Computer Engineering, University of Connecticut, Storrs, CT, USA

David Mendlovic Department of Electrical Engineering – Physical Electronics, Faculty of Engineering, Tel-Aviv University, Ramat Aviv, Israel

Vicente Micó Departamento de Óptica, Universitat de Valencia, Burjassot, Spain

Jonathan Solomon Department of Electrical Engineering – Physical Electronics, Faculty of Engineering, Tel-Aviv University, Ramat Aviv, Israel

David Sylman School of Engineering, Bar-Ilan University, Ramat-Gan, Israel

Zeev Zalevsky School of Engineering, Bar-Ilan University, Ramat-Gan, Israel

Alex Zlotnik School of Engineering, Bar-Ilan University, Ramat-Gan, Israel

Chapter 1
Theoretical Background

Alex Zlotnik, Zeev Zalevsky, David Mendlovic, Jonathan Solomon, and Bahram Javidi

1.1 Fourier Optics

1.1.1 Free Space Propagation: Fresnel and Fraunhofer Integrals

Under scalar diffraction theory assumption and assuming that work is relatively close to optical axis $\sqrt{(x-\xi)^2 + (y-\eta)^2} \ll z_0$, it is possible to write the following relationship [1]:

$$U(x,y,z_0) = \frac{\exp(jkz_0)}{j\lambda z_0} \iint U^i(\xi,\eta) \exp\left\{ j\frac{\pi}{\lambda z_0}\left[(x-\xi)^2 + (y-\eta)^2\right] \right\} d\xi d\eta. \quad (1.1)$$

This is known as the Fresnel diffraction integral. It can be calculated as a convolution between the incident field U_i and the free space propagation (FSP) quadratic phase function.

In certain limiting cases, Fresnel diffraction formula can be simplified to yield Fraunhofer diffraction integral. If the diffraction is observed on a very remote plane, the quadratic phase factor inside the integral of (1.1) can be omitted, provided that the following condition is fulfilled:

$$\frac{\pi}{\lambda z_0}\left(\xi^2 + \eta^2\right)_{max} = \pi \quad \Rightarrow \quad z_0 = \frac{D^2}{\lambda}.$$

Z. Zalevsky (✉)
School of Engineering, Bar-Ilan University, Ramat-Gan, Israel
e-mail: zalevsz@macs.biu.ac.il

Z. Zalevsky (ed.), *Super-Resolved Imaging: Geometrical and Diffraction Approaches*,
SpringerBriefs in Physics, DOI 10.1007/978-1-4614-0833-8_1,
© Springer Science+Business Media, LLC 2011

Fig. 1.1 Imaging system
consists of lens with focal
length f; (ξ, η) is the object
plane, and (x, y) is the image
plane

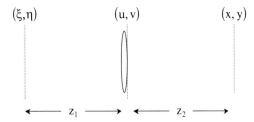

The obtained result is then:

$$U(x, y, z_0) = \frac{\exp(jkz_0)}{j\lambda z_0} \exp\left[j\frac{\pi}{\lambda z_0}\left(x^2 + y^2\right)\right]$$
$$\times \iint U^i(\xi, \eta) \exp\left[-j\frac{\pi}{\lambda z_0}(x\xi + y\eta)\right] d\xi d\eta. \qquad (1.2)$$

1.1.2 Imaging System

In this section, a simple imaging system consisting of a single thin lens of a finite aperture $P(u, v)$ and a focal length f is briefly analyzed. This system images a planar object in the (ξ, η) plane into a (x, y) image plane, while a monochromatic illumination is assumed (see Fig. 1.1).

1.1.2.1 Coherent Illumination

The output field $U_{\text{image}}(x, y)$ is related to input $U_{\text{object}}(\xi, \eta)$ through a superposition integral:

$$U_{\text{image}}(x, y) = \iint U_{\text{object}}(\xi, \eta) h(x, y; \xi, \eta) d\xi d\eta, \qquad (1.3)$$

where $h(\cdot;\cdot)$ is the amplitude at image coordinates (x, y) in response to a point – source object at (ξ, η), and is given by [1]:

$$h(x, y; \xi, \eta) = \frac{1}{\lambda^2 z_1 z_2} \exp\left[i\frac{\pi}{\lambda(z_2 - f)}\left(x^2 + y^2\right)\right]$$
$$\times \iint P(u, v) \exp\left\{-i\frac{2\pi}{\lambda z_2}[(x - M\xi)u + (y - M\eta)v]\right\} du dv, \qquad (1.4)$$

where M is the magnification, $M = -z_2/z_1$; z_1, z_2, and f obey the relation:

$$\frac{1}{z_1} + \frac{1}{z_2} = \frac{1}{f}. \tag{1.5}$$

After several simple coordinate transformations, one can obtain a convolution relationship:

$$U_{\text{image}}(x, y) = \tilde{h}(x, y) \otimes U_{\text{g}}(x, y), \tag{1.6}$$

where U_{g} is the geometrical optics prediction of the image; $\tilde{h}(x, y)$ is the point-spread function with a quadratic phase factor omitted.

1.1.2.2   Incoherent Illumination

Imaging systems using spatially incoherent illumination are linear in intensity [1] and obey the intensity convolution integral:

$$I_{\text{image}}(x, y) = \kappa \cdot |h(x, y)|^2 \otimes I_{\text{g}}(x, y), \tag{1.7}$$

where κ is a constant; $I_{\text{image}}(x, y)$ and $I_{\text{g}}(x, y)$ are intensities of $U_{\text{image}}(x, y)$ and $U_{\text{g}}(x, y)$, respectively.

1.2   Diffraction Resolution Limitation

Let us assume that we have an optical system that relies on a lens with a focal length f and aperture D. If such a system stares on a scene located at a distance of R from the sensor $(R \gg f)$, the viewed resolution in the image plane is limited by diffraction:

$$h(r) \propto \left| \frac{J_1 \left(\pi r / \lambda F_\# \right)}{r / \lambda F_\#} \right|^2 \tag{1.8}$$

and therefore equals to $1.22 \lambda F_\#$, where r is the radial coordinate in the focal plane $r = \sqrt{x^2 + y^2}$, λ is the wavelength, and $F_\#$ is the F-number of the imaging system $F_\# = f/D$.

By translating the resolution bound to the object plane, the smallest detail possibly viewed is of the size:

$$(\delta r)_{\text{diff}} = 1.22 \frac{\lambda}{D} R. \tag{1.9}$$

1.3   Geometrical Resolution Limitation

However, modern optical system are digital and contain some form of an electronic sensor. The sensor has nonzero pixels, having a size of Δd. The pixel size provides the "geometrical resolution" bound. This limitation expressed in the object plane yields:

$$(\delta x)_{\mathrm{g}} = \frac{\Delta d}{f} R. \tag{1.10}$$

In most cases, $\Delta d > 1.22(\lambda f / D)$, and the geometrical resolution is the bottleneck, in the optical system.

1.3.1   The Effects of Sampling by CCD (Pixel Shape and Aliasing)

Let us assume that an image is received on the CCD plane. The CCD samples the image with finite pixels having a defined pitch. Let us denote the distance between each pixel as Δx and the width of each pixel as Δd. Sampling the image creates replicas of the continuous image spectrum in the frequency domain. These replicas are spaced at a constant offset in the spectrum, which is proportional to the resolution of the CCD, $\Delta v = 1/\Delta x$.

Therefore, sampling the physical image by the CCD is equivalent to [2]:

(a) Convolving it with a rect function (a rectangular window) with a width equal to the size of a single CCD pixel. The latter simulates the effect of the nonzero pixel size.
(b) Multiplying the input by a comb function $\sum_m \delta(x - m\Delta x)$.

In the frequency plane, this is equivalent to:

(a) Multiplying the original's input spectrum by a sinc function $(\sin c(x) = (\sin(\pi x)/\pi x))$ with a width of $2/\Delta d$
(b) Convolving the result with a train of Dirac functions (due to the pixel spacing) $\sum_n \delta(v - (n/\Delta x))$

If the distance between the replicas is not sufficient, the replicas overlap. As a result, the image is corrupted. Figure 1.2a presents an input spectrum, and the aliased corrupted spectrum is shown in Fig. 1.2b.

Aliasing occurs when the image's resolution is more than half of that of the CCD (Nyquist sampling rate). Image resolution measured on the CCD plane is denoted as $\Delta v_{\mathrm{image}}$. In mathematical terms, aliasing occurs when $2\Delta v_{\mathrm{image}} > \Delta v$. Diffraction effects have been neglected as it is assumed that geometrical resolution bound is dominant.

Fig. 1.2 (**a**) Output image spectrum before being sampled by CCD. (**b**) Output image spectrum after being sampled by CCD. The image was taken from: J. Solomon, Z. Zalevsky and D. Mendlovic, "Geometrical Super Resolution by Code Division Multiplexing," Appl. Opt. 44, 32–40 (2005)

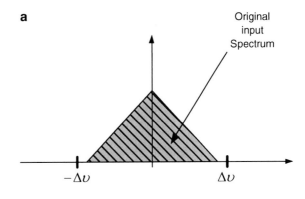

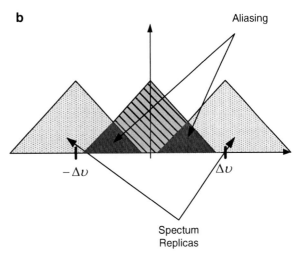

1.4   Super Resolution Explained by Degrees of Freedom Number

The possibility for super resolution is often explained by the notion of degrees of freedom (DoF) invariance of a given optical system. Other term describing the same is the information capacity of the optical system. That is the number of degrees of freedom (DoF) number the system could pass through is constant and equal to information capacity [3]:

$$N = (1 + 2L_xB_x)(1 + 2L_yB_y)(1 + 2L_zB_z)$$
$$\times (1 + 2L_TB_T)\log(1 + \text{SNR}), \tag{1.11}$$

where $L_x\,L_y$ is the field of view, L_z is the depth of field, $B_x\,B_y\,B_z$ is the spatial bandwidth in x, y, z dimensions; L_T is the observation interval and B_T is the temporal bandwidth. SNR is the signal-to-noise ratio. A priori knowledge of object properties makes possible to code the object in order to pass it through optical system inferior in certain DoF and superior in others. An example for such coding is

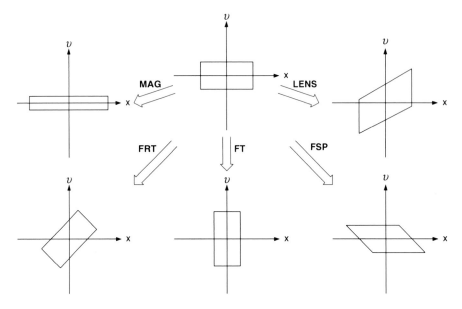

Fig. 1.3 Wigner properties

object space bandwidth (SW) shaping in Wigner space [4]. Space bandwidth product is the lateral DoF from (1.11). For 1-D signal, it is defined as

$$SW = \Delta x \Delta v, \tag{1.12}$$

Δx is the area where the signal $u(x)$ is essentially nonzero and Δv is the size of the frequency where the spectrum of $u(x)$ is essentially nonzero.

1.4.1   Wigner Transform

A Wigner chart is a wave-optical generalization of the Delano diagram (ray optics $Y\bar{Y}$ diagram). Its definition is:

$$W(x, v) = \int_{-\infty}^{\infty} u\left(x + \frac{x'}{2}\right) u^*\left(x - \frac{x'}{2}\right) \exp(-2\pi i v x') dx', \tag{1.13}$$

where $u(x)$ is the complex amplitude and v is the spatial frequency. Apparently, a Wigner chart presents the spatial and spectral information simultaneously. It doubles the number of dimensions; thus, a one-dimensional (1-D) object has a two-dimensional (2-D) Wigner chart. Figure 1.3 shows the effects of elementary optical modules, such as magnification (MAG), a lens (LENS), FSP, and Fourier transform (FT) or fractional Fourier transform (FRT), on the Wigner chart of a signal [5–8].

The definition of SW was generalized by the use of the ensemble average of the Wigner chart that is due to a set of signals that may enter the optical system. There instead of being a pure number, $SW(x, v)$ was a binary function of two variables (referring to as 1-D object) with the following definition:

$$SW_B(x, v) = \begin{cases} 1 & f\langle W(x, v) \rangle > W_{thresh}, \\ 0 & \text{otherwise.} \end{cases} \quad (1.14)$$

The area of $\lim_{x \to \infty} SW(x, v)$ indicates in fact the number of DoF; e.g., if δx denotes the spatial resolution and δv is the spectral resolution, then $\delta x = 1/\Delta v$, $\delta v = 1/\Delta x$, and the number of degrees of freedom (DoF) N is:

$$N = \frac{\Delta x}{\Delta \delta} = \frac{\Delta v}{\delta v} = \Delta x \cdot \Delta v. \quad (1.15)$$

For a given optical system whose SW acceptance capabilities are denoted by $SWY_v(x, v)$ and a given input signal whose existing SW is denoted by $SWI_v(x, v)$, a necessary condition for transmitting the whole signal without information loss is:

$$SWI_v(x, v) \subseteq SWY_v(x, v). \quad (1.16)$$

If the transmission is lossless, then the following condition takes place:

$$N_{signal} \leq N_{system}. \quad (1.17)$$

1.5 Inverse Problem Statement of Super Resolution

Achieving either geometrical or diffraction super resolution can be formulated as solving inverse problem.

Inverse problem is stated in the following manner: An image is known on a certain grid. One wishes to restore image values on a finer grid. The image is related to high-resolution unknown object through blurring, sampling, and addition of noise. The blurring is assumed to be a spatially invariant operator. It is possible to write the following discrete relationship (on a fine grid):

$$y[m, n] = g[m, n] * u[m, n] = \sum_{k=0}^{R-1} \sum_{l-0}^{R-1} g[k + m, l + n]u[k, l], \quad (1.18)$$

where $g[\ldots]$ is the blurring matrix, $u[\ldots]$ is the high-resolution object to be restored.

It is convenient to represent 2-D images as column-wise concatenated vectors and the blurring operator as a matrix. The original and blurred images are therefore assumed to be related by a compact set of linear equations:

$$\underline{\underline{A}}x = \underline{b}. \quad (1.19)$$

In order to be defined as well-posed and to have a unique solution, it must uphold the following three conditions: existence, stability, and uniqueness [9]. If some of the conditions do not hold, then the problem is ill-posed, and there may not be a solution, or it may not be unique. Furthermore, since this solution does not uphold all three conditions mentioned above, the additive noise prevent us from converging to real solution. Likewise, since there are more unknowns than equations, the solution is not unique. Finally, a small change in one of the variables would affect the solution of the problem so that the stability of the solution would be very low.

One possible direction for the above-mentioned problems is to use the pseudoinverse matrix that is obtained by a reduction of the least square error. Techniques dealing with least square error reduction [10] involve recursive least square error (RLS) [11] and recursive total least square error (RTLS) [12]. A more sophisticated method to reduce least square errors recursively uses regularization [13]. This method which succeeds to overcome noise contains Tikhonov's regularization component. This component is designed such that for problems without noise it will be possible to reduce it so that the real solution will be approached, while for images with noise this positive addition will yet yield an optimal solution [14].

There is a set of other regularization methods that uses prior knowledge of the system regarding the statistical properties of the blurring problem. This set of methods is called stochastic reconstruction methods. In this set of methods, reconstruction of a super resolution image is a statistical re-evaluation problem, where all quantities are modeled by their probability functions. One way to reconstruct is by applying the maximum a posteriori (MAP) where the super resolved image may be obtained by looking for the maximum of the conditional probability distribution whose estimation is done by the Markov random field (MRF) in different ways, enabling the addition of a priori constraints into the solution [15, 16]. Another solution known as maximum likelihood (ML) is actually a particular case of MAP, where the required image is obtained by the ML estimator which does not need any a priori knowledge [17].

A different approach named projection onto convex sets (POCS) assumes a number of prior demands of the required solution. For each such demand, an operator is defined that projects a dot in the field of the super resolution image onto a field fulfilling the constraint. Such an iterative process of operator activation causes the solution to converge fulfilling all the constraints and even may avoid guessing the first solution either by using the time domain [18] or by using the frequency domain [19]. Following that another interesting direction for solving the blurring problem by iteration is via using the iterated back-projection (IBP) method [20].

References

1. Goodman, J.W.: Introduction to Fourier Optics, Chap. 3, 2nd edn. McGray-Hill, New York (1996)
2. Oppenheim, A.V., Willsky, A.S., Hamid, S.: Signals and Systems, Chap. 7, 2nd edn. Prentice-Hall, London (1996).
3. Cox, I.J., Shappard, C.J.R.: Information capacity and resolution in an optical system. J. Opt. Soc. Am. A **3**, 1152–1158 (1986)
4. Lohmann, A.W., Dorsch, R.G., Mendlovic, D., Zalevsky, Z., Ferreira, C.: Space-bandwidth product of optical signals and systems. J. Opt. Soc. Am. A **13**(3), 470–473 (1996)
5. Zalevsky, Z., Mendelovic, D., Lohmann, A.W.: Understanding superresolution in Wigner space. J. Opt. Soc. Am. **17**(12), 2422–2430 (2000)
6. Mendelovic, D., Lohmann, A.W.: Space-bandwidth product adaptation and its application to superresolution: fundamentals. J. Opt. Soc. Am. A **14**, 558–562 (1997)
7. Mendelovic, D., Lohmann, A.W., Zalevsky, Z.: Space-bandwidth product adaptation and its application to superresolution: examples. J. Opt. Soc. Am. A **14**, 563–567 (1997)
8. Lohmann, A.W.: Image rotation, Wigner rotation and the fractional Fourier transform. J. Opt. Soc. Am. A **10**, 2181–2186 (1993)
9. Hadamard, J.: Lectures on Cauchy's Problem in Linear Partial Differential Equation. Dover, New York (1923)
10. Tekalp, A.M., Ozkan, M.K., Sezan, M.I.: High-resolution image reconstruction from lower-resolution image sequences and space varying image restoration. In: Proceedings of the IEEE International Conference on Acoustics, Speech and Signal Processing (ICASSP), vol. 3, pp. 169–172 (1992)
11. Kim, S.P., Bose, N.K., Valenzuela, H.M.: Recursive reconstruction of high resolution image from noisy undersampled multiframes. IEEE Trans. Acoust. Speech Signal. Process. **38**(6), 1013–1027 (1990)
12. Bose, N.K., Kim, H.C., Valenzuela, H.M.: Recursive total least squares algorithm for image reconstruction from noisy, undersampled frames. Multidim. Syst. Signal Process. **4**(3), 253–268 (1993)
13. Kim, S.P., Su, W.Y.: Recursive high-resolution reconstruction of blurred multiframe images. IEEE Trans. Image Process. **2**(4), 534–539 (1993)
14. Tikhonov, A., Arsenin, V.Y.: Solutions of Ill-posed Problems. V.H Winston and Sons, Washington, DC (1977)
15. Cheeseman, P., Kanefsky, B., Kraft, R., Stutz, J., Hanson, R.: Super-resolved surface reconstruction from multiple images. In: Heidbreder, G.R. (ed.) Maximum Entropy and Bayesian Methods, pp. 293–308. Kluwer Academic Publishers, Dordrecht, The Netherlands (1996)
16. Hardie, R.C., Barnard, K.J., Armstrong, E.E.: Joint MAP registration and high-resolution image estimation using a sequence of undersampled images. IEEE Trans. Image Process. **6**(12), 1621–1633 (1997)
17. Tom, B.C., Katsaggelos, A.K.: Reconstruction of a high-resolution image from multiple-degraded misregistered low-resolution images. Vis. Commun. Image Process. SPIE **2308**(2), 971–981 (1994)
18. Stark, H., Oskoui, P.: High-resolution image recovery from image-plane arrays using convex projections. J. Opt. Soc. Am. A**6**, 1715–1726 (1989)
19. Wheeler, F.W., Hoctor, R.T., Barrett, E.B.: Super-resolution image synthesis using projections onto convex sets in the frequency domain. In: IS&T/SPIE Symposium on Electronic Imaging, Conference on Computational Imaging, vol. 5674, pp. 479–490 (2005)
20. Irani, M., Peleg, S.: Motion analysis for image enhancement: resolution, occlusion, and transparency. J. Vis. Commun. Image R. **4**(4), 324–335 (1993)

Chapter 2
Super Resolution Methods Implementing Diffractive Masks Having a Certain Degree of Periodicity

Alex Zlotnik, Zeev Zalevsky, Vicente Micó, Javier García, and Bahram Javidi

2.1 Single Snap-Shot Double Field Optical Zoom

2.1.1 Introduction

This section presents an approach that provides super resolved imaging at the center of the field of view and yet allows to see the remaining of the original field of view with original resolution. This operation resembles optical zooming while the zoomed and the nonzoomed images are obtained simultaneously. This is obtained by taking a single snap-shot and using a single imaging lens. The technique utilizes a special static/still coding element and a postprocessing algorithmic, without any mechanical movements.

Optical zooming is basically a super resolution technique since its purpose is to obtain resolution higher than provided by the imaging system (prior to zooming). The physical restrictions that limit the spatial resolution of an imaging system are either the size of aperture of the imaging lens or the geometrical parameters of the detection array such as its pitch and fill factor. Eventually, the hardest limitation prevails.

The common optical realization of optical zoom includes several lenses and a mechanical mechanism as in ref. [1]. Other principles do not include mechanical movements but rather other time adaptive concepts allowing variation of the overall focal length of the lens. In the literatures [2–13], one may see an example of several works dealing with zooming lenses. Thus, basically the zooming operation is actually the increase in focal length of the imaging module providing

Z. Zalevsky (✉)
School of Engineering, Bar-Ilan University, Ramat-Gan, Israel
e-mail: zalevsz@macs.biu.ac.il

Z. Zalevsky (ed.), *Super-Resolved Imaging: Geometrical and Diffraction Approaches*, SpringerBriefs in Physics, DOI 10.1007/978-1-4614-0833-8_2, © Springer Science+Business Media, LLC 2011

smaller foot print of each pixel in the detector, on top of the object. The spatial resolution improvement in the center of the field of view during the zooming process is obtained since the foot print of each pixel on the object equals to $\Delta x R/F$, where Δx is the pitch of the pixels of the detector, F is the focal length, and R is the distance from the object.

Thus, the regular optical zooming operation has two major disadvantages. The first one is that the increase of the focal length, for instance, by a factor of 3, while preserving the F-number will result in increase in the volume of the imaging module by a factor of $3^3 = 27$. It means more weight and less reliability (due to the mechanical mechanism). The second disadvantage is that the zoomed and the nonzoomed images are not obtained simultaneously, and the resolution improvement in the central part of the field of view comes on the expense of decreasing the field of view. Note that the resolution improvement in the center of the field of view is not due to the increase of the focal length F but is rather due to the generation of smaller effective pixels in that spatial region. That is reduction of Δx by the same factor in which the F-number has been increased.

In the approach reviewed in this section, the zoomed and nonzoomed images are obtained *simultaneously in a single snap-shot*. It should be noted that the resolution improvement obtained in the central part of the field of view [14] follows the idea presented in ref. [15]. However, the idea in ref. [15] shows how to obtain the resolution improvement, but here it is shown how to obtain this improvement *without* sacrificing the field of view, i.e., also obtaining the nonzoomed resolution in the remaining part of the field of view. Note that having an improved resolution in the central part of the field of view and simultaneously preserving the original nonzoomed resolution in the outer parts yield more spatially resolved points than the number of pixels in the detector array. Such an outcome is made possible by a trade-off payment in the dynamic range of the captured image.

The operation principle is based on the followings: the image resolution obtained using a common single lens is higher in the center of the field of view and degrades toward the periphery. Usage of this property is essential for the proposed operation principle. This is because the surface where a perfect image is obtained is rather a sphere than a plane. The optical limit for the resolution obtained in the center is proportional to $\lambda F/D$ (where λ is the wavelength, F is the focal length, and D is the aperture of the lens). For many detectors, this resolution limit is much less restrictive and harder to reach in comparison to the restriction due to the sampling pitch of the detector. Consequently, in such cases, the detector is forced to get a poor image quality. In our technique, the optics provides, in the center of the field of view, an optical resolution that is limited by the diffraction. In the remaining part of the field of view, the optics provides a resolution limit which equals to the detector's sampling pitch. In this manner by exploiting the aliasing effect due to sampling of the detector and by performing some digital postprocessing result in a super resolved image. The image has a diffraction limited resolution at the center region of the field of view and yet preserves the original geometrical resolution at its outer parts.

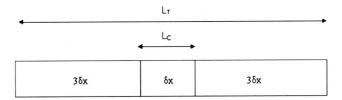

Fig. 2.1 One-dimensional object. The minimal details in the central part are three times finer, than those in the periphery. The image was taken from: Z. Zalevsky and A. Zlotnik, "Single Snap-Shot Double Field Optical Zoom," Opt. Exp. 13, 9858–9868 (2005)

2.1.2 Theory

2.1.2.1 Preliminary

For the sake of simplicity, the analysis of the method is one-dimensional (1-D). A two-dimensional deduction is straight forward. Let us take a 1-D positive object $s(x)$ (see Fig. 2.1). Its spatial support is denoted by L_T. This object has minimal resolution detail denoted by δ_x in its central L_C part. In the following mathematical analysis, L_C is taken to be 1/6 of L_T, although other ratio can be chosen. The finest optically resolved detail in the remaining periphery is three times larger, $3\delta_x$, than that equals to the geometrical limitation of the pitch of the sampling detection array (see Fig. 2.1). This limitation is determined by the optics and exists prior to the digital sampling performed by the detection array.

One wishes to image this object using an ideal aberration-free optical system with a magnification factor of 1. The image is captured using a camera with a pixel pitch of $3\delta_x$, while pixels are assumed to represent an ideal spatial Dirac impulse train. The proposed method enables resolving details with a high resolution in the central part, in spite of the larger pitch, without decrease in the field of view. Therefore, an optical zooming of ×3 in the central 1/6 field of view is obtained, while having simultaneously the ×1 resolution (without zooming) in the other 5/6 field of view. All of this is obtained from a single optically coded and then digitally processed image. The penalty is the introduction of some noise in the obtained image. The optical coding involves insertion of a certain spatial coding grating in the entrance pupil plane of the imaging lens. The super resolving approach that increases the resolution in the central 1/6 part of the field of view is based upon the approach presented in ref. [15]. The investigated case here deals with coherent illumination, although extension into noncoherent case is straight forward as described in ref. [15].

Note that the geometrical super resolution method described in ref. [15] is equivalent to the realization of an optical zoom in the central part of the field of view since the footprint seen, in the super resolved image, over the observed object equals to: $(R/F)(\Delta x/\kappa)$ where R is the distance between the camera and the object, F is the focal length, Δx is the pitch of the pixels of the camera ($\Delta x = 3\delta$), and κ is the geometrical super resolution factor (the case of $\kappa = 3$ is assumed hereafter). In case that optical zoom of factor κ is performed, the focal length is changed to κF and thus the footprint equals to $(R/(\kappa F))(\Delta x)$. It is easily seen that both expressions are identical. Thus, in ref. [15], it is shown that how without changing the focal length it is possible to perform optical zooming, which is actually done by performing geometrical super resolution. However, the condition for the operation of the approach presented in ref. [15] is that the input object occupies no more than $1/\kappa$ of the field of view.

2.1.2.2 Mathematical General Description

$S(v)$ denotes the Fourier transform of the object $s(x)$, with the spatial frequency coordinate, v, belonging to the spectral range of $\in [-v_{max}, v_{max}]$, where v_{max} is the maximal spatial frequency of the object. It is inversely related to the spatial resolution δ_x. The Fourier content is virtually divided into three equal regions:

(a) Left third $S_{-1}(v)$ with $v \in [-v_{max}, -1/3v_{max}]$
(b) Central third $S_0(v)$ with $v \in [-1/3v_{max}, 1/3v_{max}]$
(c) Right third $S_1(v)$ with $v \in [1/3v_{max}, v_{max}]$.

These spectral components are multiplied by the spatial grating so that a certain degree of orthogonality between the components is created. The coding grating mask also consists of three regions:

(a) Left third $G_{-1}(v)$ with $v \in [-v_{max}, -1/3v_{max}]$
(b) Central third $G_0(v)$ with $v \in [-1/3v_{max}, 1/3v_{max}]$
(c) Right third $G_1(v)$ with $v \in [1/3v_{max}, v_{max}]$

The chosen mask fulfills the orthogonality condition of:

$$G_l(v)G_k(v) = \delta[l, k], \tag{2.1}$$

where $\delta[l, k]$ is the Kronicker delta function. When the image is under-sampled by the detector, an aliasing effect takes place. In fact, the aliasing is essentially a folding of $S_{-1}(v)$ and $S_1(v)$ into a central third of the spectrum. Therefore, the spectrum of the captured image equals to:

$$I(v) = \sum_{k=-1}^{1} S_k(v) \times G_k(v) \quad v \in \left[\frac{-1}{3v_{max}}, \frac{1}{3v_{max}}\right]. \tag{2.2}$$

To improve the clarity of this presentation, let us now briefly recall the derivation made in ref. [15]. Let us examine a simple situation, in which the goal is improving the resolution by a factor of 3. Assuming an ideal CCD in which the pixels are indefinitely small and are placed at a distance of Δx from one another (according to Fig. 2.1, $\Delta x = 3\delta_x$). Next it is shown that when one pixel is willing to sacrifice 1/3 of the field of view, the other can obtain an improvement in the resolution of that central 1/3 of the field of view by a factor of 3 (without increasing the focal length by a factor of 3). In the case of ideal sampling, the sampling function of the CCD [denoted as CCD(x)] is modeled as an infinite train of impulses:

$$CCD(x) = \sum_{n=-\infty}^{\infty} \delta(x - n\Delta x). \tag{2.3}$$

As previously mentioned, the coding mask [denoted as $CDM\tilde{A}(v)$] is divided into three subfunctions as follows:

$$CDM\tilde{A}(v) = \sum_{n=-1}^{1} G_n(v - n\Delta v). \tag{2.4}$$

The CDMA mask is multiplied by the Fourier plane with the spectrum of the input signal $s(x)$ [denoted as $S(v)$]. This is obtained by positioning the coding mask in the coherent transfer function (CTF) plane of the imaging lens. In the coherent case, in the CTF plane, a Fourier of the imaged object is obtained. In the noncoherent case, this position is also related to the spectrum of the imaged object.

This spatial distribution is multiplied by CCD(x), the sampling grid of the CCD, which means that it is convolved with the Fourier of the CCD grid in the spectral domain:

$$D(v) = \left[S(v) \sum_{n=-1}^{1} G_n(v - n\Delta v) \right] * \left[\sum_{n=-\infty}^{\infty} \delta\left(v - n\frac{2\pi}{\Delta x}\right) \right], \tag{2.5}$$

where * denotes the convolution operation. Since $\Delta v = 2\pi/\Delta x$, the last expression can be simplified to:

$$D(v) = \left[S(v) \sum_{n=-1}^{1} G_n(v - n\Delta v) \right] * \left[\sum_{n=-\infty}^{\infty} \delta(v - n\Delta v) \right],$$

$$= \sum_{n=-\infty}^{\infty} S(v - n\Delta v) \left[\sum_{k=-1}^{1} G_n(v - (n+k)\Delta v) \right], \tag{2.6}$$

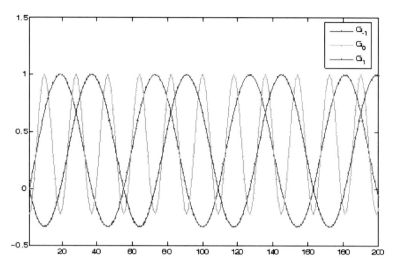

Fig. 2.2 The spatial grating positioned in CTF plane. Its three parts $G_{-1}(v)$, $G_0(v)$, and $G_1(v)$ are plotted in a folded manner. The period of $G_0(v)$ is three times smaller than that of $G_{-1}(v)$ and $G_0(v)$. The image was taken from: Z. Zalevsky and A. Zlotnik, "Single Snap-Shot Double Field Optical Zoom," Opt. Exp. 13, 9858–9868 (2005)

Image retrieval is simply achieved by Fourier transforming the grabbed output and multiplying it with the original coding mask and then downsampling:

$$R(v) = D(v)\text{CDM}\tilde{A}(v) = \left\{ \sum_{n=-\infty}^{\infty} S(v - n\Delta v)\left[\sum_{k=-1}^{1} G_n(v - (n+k)\Delta v)\right]\right\}$$
$$\left[\sum_{m=-1}^{1} G_m(v - m\Delta v)\right],$$
$$= \sum_{n=-\infty}^{\infty} S(v - n\Delta v)G_n(v - n\Delta v) = S(v)\text{CDM}\tilde{A}(v) \xrightarrow[\text{downsampling}]{} S(v), \quad (2.7)$$

Modulating the input's spectrum by multiplying with the coding mask correctly prevents data corruption due to aliasing. This insight was proven in ref. [15] and demonstrated experimentally. It indeed demonstrates super resolution, i.e., an effect equivalent to seeing an image with a zoom of ×3 without changing the focal length. But this improvement is obtained only in the central 1/3 of the field of view while the input object occupies only 1/3 of the field of view. Next it is farther proved that it is possible to obtain the super resolved image in the central field of view without the need of paying with the outer 2/3 of the field.

The grating of (2.1) is illustrated in Fig. 2.2 in a folded manner: $G_{-1}(v)$ and $G_1(v)$ are folded into a central third part of the spectrum of $G_0(v)$. As a result, $I(v)$ can be described as composed of so-called "macropixels." Each macropixel

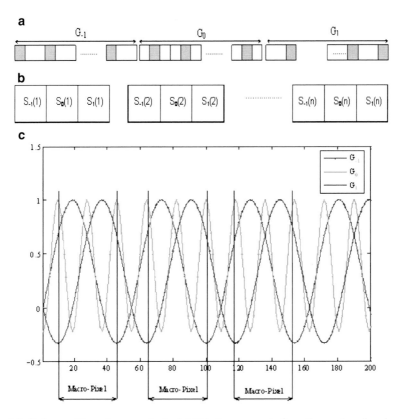

Fig. 2.3 Orthogonality and macro pixels: (**a**) This is an example for orthogonal coding: in each spectral region there is a macropixel with a certain nonzero pixel. (**b**) After aliasing, all nonzero pixels are folded in a nonoverlapping way, providing orthogonality. (**c**) Due to the real realization of the grating, the true structure is a little bit different from the theory presented in parts (**a**) and (**b**) of the figure. The image was taken from: Z. Zalevsky and A. Zlotnik, "Single Snap-Shot Double Field Optical Zoom," Opt. Exp. 13, 9858–9868 (2005)

consists of the $S_{-1}(v)$, $S_1(v)$, and $S_0(v)$ contributions (see Fig. 2.3a–c). The structure presented in Fig. 2.3a and b is the theoretical goal since it provides full and simple orthogonality condition. In reality, however, such binary-like coding grating will have finite number of harmonics. Therefore, the spectral structure of the "macropixels" will be different. However, if properly designed, it will yet remain orthogonal (when proper locations are observed) and will resemble the structure showed in Fig. 2.3c.

Next, the reconstruction algorithm for the original image is formulated. The orthogonal coding grating mask is a Dammann-like phase structure whose spatial effect is similar to replications. The mask is designed such that a different replication is generated for the high- $[G_{-1}(v)$ and $G_1(v)]$ and low-frequencies content $[G_0(v)]$ as shown in Fig. 2.4. The replications for high frequencies are 1/6 field of view apart and for low frequencies are 1/2 field of view apart.

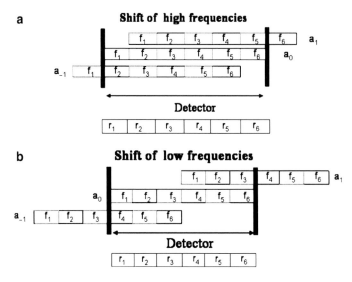

Fig. 2.4 Spatial effect of the coding mask. (**a**). Replication of high spectral content. (**b**) Replication of the low spectral content. The image was taken from: Z. Zalevsky and A. Zlotnik, "Single Snap-Shot Double Field Optical Zoom," Opt. Exp. 13, 9858–9868 (2005)

1. First, the high-frequency content $S_{-1}(v)$ and $S_1(v)$ is reconstructed by sampling $I(v)$: The spatial contents of $S_{-1}(v)$ and $S_1(v)$ occupy only a fraction of the field of view L_T. Therefore, it is possible to keep only each sixth (L_T/L_C) sample without losing information. Other samples are calculated using interpolation. Figure 2.5a illustrates the sampling grid. Note that the sampling points of $S_{-1}(v)$ and $S_1(v)$ are orthogonal. On the other hand, there is a certain noise added to the sampled high-frequency content due to $S_0(v)$. In order to minimize this noise effect, each sample value is taken to be as algebraic average in its neighborhood. Figure 2.5b shows the Fourier transform of the grating illustrated in Fig. 2.5a. As one may note, it resembles seven delta functions: the two pairs of delta functions appearing on both sides of the central delta resemble spatial derivative since each one of those two pairs contain one positive and one negative delta while small spatial shift is introduced between them. Those two pairs that make the derivative correspond to the two replications (the -1 and the 1 orders) related to the high frequencies (Fig. 2.4a). The outer two deltas correspond to the two replications (again the -1 and 1 orders) of the low frequencies (Fig. 2.4b).
2. Next, the reconstructed $S_{-1}(v)$ and $S_1(v)$ are subtracted from $I(v)$. Ideally, this leaves only the low-frequency content. It is expressed in the spatial domain as:

$$i_L(x) = (s_0 * g_0)(x) \cdot \text{rect}\left(\frac{x}{L_T}\right), \tag{2.8}$$

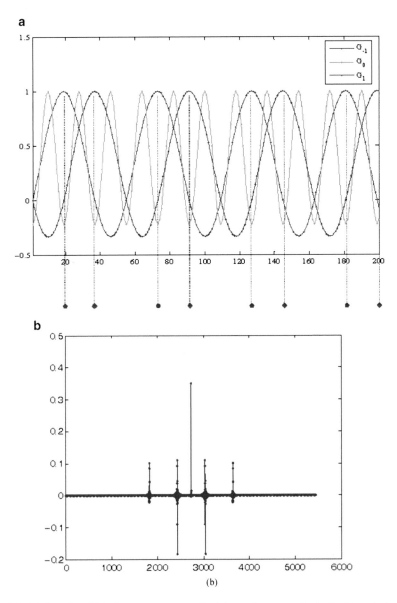

Fig. 2.5 (**a**) Sampling high-frequency content: $S_{-1}(\nu)$ samples are marked with *blue circles* and $S_1(\nu)$ samples are marked with *red diamonds*. (**b**) The Fourier transform of the grating. The image was taken from: Z. Zalevsky and A. Zlotnik, "Single Snap-Shot Double Field Optical Zoom," Opt. Exp. **13**, 9858–9868 (2005)

where s_0 and g_0 are the inverse Fourier transforms of $S_0(v)$ and $G_0(v)$, respectively, and "*" stands for convolution operation. $\text{rect}(x/L_T)$ is defined as:

$$\text{rect}\left(\frac{x}{L_T}\right) = \begin{cases} 1 & |x| \leq L_T/2, \\ 0 & \text{otherwise.} \end{cases} \tag{2.9}$$

The $g_0(x)$ is in fact consists of three Dirac impulse functions:

$$g_0(x) = \sum_{n=-1}^{1} a_n \times \delta\left(x - \frac{nL_T}{2}\right), \tag{2.10}$$

3. Now each $i_L(x)$ and $s_0(x)$ are divided into sets of six equally supported functions, denoted correspondingly as $r_j(x)$ $j = 1, \ldots, 6$ and $f_j(x)$ $j = 1, \ldots, 6$. These two sets of functions are related through six linear equations. Those equations can be well understood after observing Fig. 2.4b:

$$\begin{aligned}
r_1(x) &= a_0 f_1(x) + a_{-1} f_4(x), \\
r_2(x) &= a_0 f_2(x) + a_{-1} f_5(x), \\
r_3(x) &= a_0 f_3(x) + a_{-1} f_6(x), \\
r_4(x) &= a_0 f_4(x) + a_1 f_1(x), \\
r_5(x) &= a_0 f_5(x) + a_1 f_2(x), \\
r_6(x) &= a_0 f_6(x) + a_1 f_3(x)
\end{aligned} \tag{2.11}$$

or alternately through a 6×6 matrix:

$$\begin{bmatrix} r_1(x) \\ r_2(x) \\ r_3(x) \\ r_4(x) \\ r_5(x) \\ r_6(x) \end{bmatrix} = \begin{bmatrix} a_0 & 0 & 0 & a_{-1} & 0 & 0 \\ 0 & a_0 & 0 & 0 & a_{-1} & 0 \\ 0 & 0 & a_0 & 0 & 0 & a_{-1} \\ a_1 & 0 & 0 & a_0 & 0 & 0 \\ 0 & a_1 & 0 & 0 & a_0 & 0 \\ 0 & 0 & a_1 & 0 & 0 & a_0 \end{bmatrix} \begin{bmatrix} f_1(x) \\ f_2(x) \\ f_3(x) \\ f_4(x) \\ f_5(x) \\ f_6(x) \end{bmatrix}. \tag{2.12}$$

The equation 2.12 can be solved to obtain the set of the $f_j(x)$, which is the low frequency content of the original image information. Note that $f_i(x)$ are the original six spatial regions of $s(x)$ while $r_i(x)$ are the spatial distributions obtained in each of the six regions after generation of the replications on the CCD plane. Equations 2.11–2.12 correspond to the low-frequency shift shown in Fig. 2.4b. a_i are the coefficients with which each one of the three replication in Fig. 2.4b is multiplied.

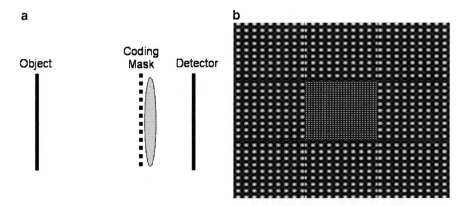

Fig. 2.6 (**a**) The experimental setup. (**b**) The coding Dammann mask that was attached to the imaging lens. The image was taken from: Z. Zalevsky and A. Zlotnik, "Single Snap-Shot Double Field Optical Zoom," Opt. Exp. 13, 9858–9868 (2005)

2.1.3 Simulation Investigation

In the experiment, it is assumed that the test object is imaged with an optical imaging system having a resolution limit in the periphery equals to the detector's array pitch. In the central part of the field of view, the optical resolution is three times larger than in the periphery. In the simulations, a Lena image is used as an object. A high-frequency 2-D barcode is planted in the center of this image. This barcode is under-sampled if its every third pixel is taken into consideration. Therefore, the central high-frequency content of the image, that is the barcode pattern, is under-sampled or low-pass filtered by a detector. To adapt the notations of Fig. 2.1, the resolution of Lena image is $3\delta_x$ while the resolution of the barcode pattern is δ_x. A grating element (the coding mask) was attached to the imaging lens (the CTF plane or the entrance pupil of the lens) as depicted in the experimental setup of Fig. 2.6a. The grating contains a different Dammann grating (see ref. [16]) in the central and outer parts of the mask as described in Fig. 2.4. The mask itself is illustrated in Fig. 2.6b.

The three regions of the grating depicted in Fig. 2.3c are merely a shifted cosine functions. In this arrangement, the high-frequency content is sampled at 1/6 of the basic sampling rate, since the spatial extent of $S_{-1}(v)$ and $S_1(v)$ is $L_C = L_T/6$. The Fourier transform of the grating is merely several impulse functions that in the spatial domain generate the six-shifted replicas of the object, as shown in Fig. 2.5b. After recovering the high-frequency content, one can solve a set of six linear equations [see (2.11) or (2.12)] in order to reconstruct the low-frequency content $S_0(v)$. Figure 2.7a presents the nonzoomed image in which the full field of view is seen. In this case, though, the central high-resolution barcode structure cannot be resolved (see Fig. 2.7a). In Fig. 2.7b, a regular optical zooming to the image of Fig. 2.7a is performed. Here, the field of view is reduced by a factor of 3 but the spatial resolution is improved by the same factor and now the central barcode structure can be resolved.

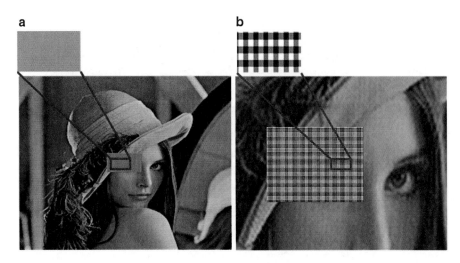

Fig. 2.7 (**a**) The nonzoomed test object used in simulations: Lena image with high-frequency two-dimensional barcode pattern at its center. (**b**) The ×3 zoomed test target where one may see the high-frequency barcode pattern. The image was taken from: Z. Zalevsky and A. Zlotnik, "Single Snap-Shot Double Field Optical Zoom," Opt. Exp. 13, 9858–9868 (2005)

Fig. 2.8 The obtained result after the digital decoding. One may see the full field of view and the zoomed highly resolved barcode pattern in the center of the field of view. The image was taken from: Z. Zalevsky and A. Zlotnik, "Single Snap-Shot Double Field Optical Zoom," Opt. Exp. 13, 9858–9868 (2005)

In the final stage, the captured image is postprocessed. The resulted image is shown in Fig. 2.8. This image proves the concept presented here: The high-frequency central field of view ($\times$3 optical zoom) is retrieved along with the nonzoomed remaining field of view. Obviously, the 6×6 spatial blocks seen on the reconstructed image in Fig. 2.8 can be removed by proper image processing and enhancement that was not applied on the obtained image.

2.2   Full Field of View Super Resolution Imaging Based on Two cStatic Gratings and White Light Illumination

2.2.1   Introduction

The usage of two static gratings for obtaining super resolved imaging dates back to the work by Bachl and Lukosz in 1967 [17, 18]. Later on, it was expanded and tested by other researchers [19–21].The Lukosz method is to place two static gratings in one out of two possible configurations: one grating before the object and the second one before the image or one grating between the object and the image and the second one after the image plane. Then, a super resolved imaging is obtained while payment in the field of view is assumed. In order to have super resolution, the two fixed gratings create ghost images which limit the field of view around the region of interest. Except the reduction in the field of view, the concept is applicable in a simple way since the two gratings are static. However, the discussed system had a magnification of one and therefore it had one major problem: one out of the two gratings is not positioned between the object and the image. This means that for the case of placing the first grating after the object, a second imaging lens is required in order to image the second grating (which is positioned after the intermediate image plane) on the output plane. In order to do that, the second imaging lens must provide a resolution as high as the resolution that one wish to extract and therefore the super resolution performed to the first lens seems to be not useful.

In this section, a modification to the super resolution approach with two main novelties is presented [22]. First, a polychromatic illumination is used instead of monochromatic one. Since the position of each ghost image is wavelength dependent (due to the gratings), the various images are averaged. Therefore, no limitation on a restricted field of view is required any longer. The possibility not to limit intentionally the field of view is very important. Practically it eliminates the need in the intermediate imagery. The payment will be done in the dynamic range required from the sensor. The second improvement is that the imaging system constructed has large magnification ratio and therefore the second grating is magnified as well to match the first grating. Due to the difference in magnification in comparison to the magnification of first one, the spatial period of the second grating is also very large and the addition of the second imaging lens in order to image it to the output plane does not require a high-resolution lens. Therefore, the gratings perform super resolution only on the first imaging lens and the setup therefore is much more effective.

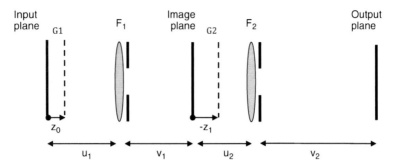

Fig. 2.9 Sketch of the proposed experimental setup

2.2.2 *Mathematical Analysis*

In the following, it is proved that indeed when a polychromatic illumination is used instead of a monochromatic one, the approach of two fixed gratings can provide super resolved imaging without paying with the field of view. The trade off in this case will be the dynamic range since the undesired replications will be averaged into a uniform intensity distribution (Because the various replications do not fall on the same spatial position, their summation is equivalent to spatial averaging of the product between the image of the object and the spectral distribution of the illumination. Such an averaging approximately yields a constant in case of large number of spatially dispersed replications). Figure 2.9 depicts the sketch of the proposed optical setup. The setup includes two cascaded imaging modules. The first has a magnification of $M_1 = u_1/v_1$ and the second of $M_2 = u_2/v_2$. The fixed gratings G_1 and G_2 are positioned at distances of z_0 and z_1 from the input and the intermediate image planes, respectively. The focal lengths of the two imaging lenses are F_1 and F_2, respectively. The values of v_1, u_1, F_1 as well as v_2, u_2, F_2 fulfill the imaging relation. Each one of the two imaging lenses has finite aperture determining its limits of spatial resolution.

In the two fixed gratings approach, the first grating is used as an encoding function (that encodes the spatial information of the input object and allows its transmission through the band limited aperture of the imaging lens) while the second is used as a decoder (that reconstructs the encoded information and produces the super resolved image). Both must have identical spatial distributions except for a scaling factor that depends on the ratio between the magnifications of the two parts of the optical configuration of Fig. 2.9. The ratio between M_2 and M_1 is large such that the spatial period of the required grating G_2 will be very large and will not be deformed by the cut-off frequency of the imaging lens of the second part of the configuration of Fig. 2.9.

In order to prove the effect of polychromatic illumination, due to reasons of simplicity for the mathematical validation, it is assumed that $M_1 = M_2 = 1$, i.e., $v_1 = u_1 = 2F_1$ and $v_2 = u_2 = 2F_2$ as well as $F_1 = F_2 = F$. The mathematical derivation is provided in ref. [21]. In that paper, an optical setup with assumptions similar to presented here is mathematically analyzed. The analysis is based upon basic Fourier optics relations while the outline for the formulation is as follows: The input field distribution is a free space propagated at a distance of z_0 and multiplied by the first grating G_1. Then it is virtually propagated backwards by a free space distance of $-z_0$ in order to reflect the effect of this grating over the input plane. This result is Fourier transformed and multiplied by a rectangular function rect$(\Delta\mu/\lambda 2F)$ where $\Delta\mu$ represents the lateral extent of the aperture of the imaging lens. The result is inverse Fourier transformed to reach the image plane. The distribution there is free space propagated a distance of $-z_1$ and multiplied by the grating G_2 and then propagated backwards by a free space distance of z_1 in order to reflect the grating on the image plane which is imaged with a magnification of 1 (in the simplified assumption) to the output plane. The field distribution obtained in the image or in the output plane, after all those mathematical procedures, is given as follows:

$$
u_0(x, z = 4F) = \sum_m \sum_n A_m B_n \int_{-\infty}^{\infty} \tilde{u}_0(v) \mathrm{rect}\left(\frac{v + mv_0}{\Delta\mu/\lambda 2F}\right) \times \exp\left[2\pi i \left(x(mv_0 + nv_1)\right.\right.
$$
$$
\left.\left. + v(z_0 \lambda mv_0 - z_1 \lambda nv_1) + \frac{z_0 \lambda m^2 v_0^2}{2} - \frac{z_1 \lambda n^2 v_1^2}{2} - z_1 \lambda mn v_0 v_1\right)\right] \times \exp[2\pi i xv] \mathrm{d}v,
$$

$$(2.13)$$

where v_0 and v_1 are the fundamental frequencies of the gratings G_1 and G_2, respectively. A_m and B_n are the Fourier series coefficients of those gratings, respectively. $\tilde{u}_0(v)$ is the Fourier transform of the high-resolution input field distribution. n and m are integers and λ is the optical wavelength. v is the spectral coordinate. In this simplified configuration, the axial location of $z = 4F$ is the position of the image plane which is basically also the output plane since the effect of the grating G_2 that appears after the image plane was already taken into account (i.e., reflected to the output plane).

The physical meaning of (2.13) is explained as follows: basically, it is an inverse Fourier transform of the Fourier of the input field distribution $\tilde{u}_0(v)$ multiplied by a synthetic aperture and an additional phase term. Due to the summation over the index m, the spectrum of the input field $\tilde{u}_0(v)$ is actually multiplied by a synthetic aperture which is wider than the original aperture that is set by the dimensions of the imaging lens. The rect expression is synthetically enlarged due to its replications and following the summation over the index "m". Therefore, more spatial frequencies can pass through the output image which contains spatial resolution that is equivalent to the one confined within the input field distribution. However, the expression of (2.13) contains also an undesired phase term of:

$$\exp\left[2\pi i\left(x(mv_0 + nv_1) + v(z_0\lambda mv_0 - z_1\lambda nv_1) + \frac{z_0\lambda m^2 v_0{}^2}{2} - \frac{z_1\lambda n^2 v_1{}^2}{2} - z_1\lambda mnv_0 v_1\right)\right].$$

In order to have true super resolution, this term is assumed to be a constant that does not affect the inverse Fourier transform integral. Choosing $z_0 = -z_1$ and $v_0 = v_1 = \Delta\mu/\lambda 2F$ (two identical gratings) which yields:

$$u_0(x, z = 4F) = \sum_m \sum_n A_m B_n \int_{-\infty}^{\infty} \tilde{u}_0(v)\text{rect}\left(\frac{v + mv_0}{\Delta\mu/\lambda 2F}\right).$$
$$\exp\left[2\pi i\left(x(v + v_0(m + n)) + v(z_0\lambda(m + n)v_0) + \frac{z_0\lambda v_0{}^2}{2}(m + n)^2\right)\right]dv.$$
$$(2.14)$$

Note that the value of z_0 and z_1 are measured according to the notations of Fig. 2.9. This means that choosing $z_0 = -z_1$ means that either G_1 and G_2 are in front of the input and image planes, respectively, or both are after those planes. For $n = -m$, super resolution is obtained since then:

$$u_0(x, z = 4F) = \int_{-\infty}^{\infty} \tilde{u}_0(v)\left[\sum_m A_m B_{-m}\text{rect}\left(\frac{v + mv_0}{\Delta\mu/\lambda 2F}\right)\right]\exp[2\pi ixv]dv. \quad (2.15)$$

The expression in (2.15) is exactly the proof of super resolution since the spectrum of the input field distribution $\tilde{u}_0(v)$ is multiplied by an extended synthetic aperture (the term in brackets) allowing transmission of higher spatial frequencies and therefore reconstruction of the output field $u_0(x, z = 4F)$ containing smaller spatial details.

The meaning of choosing $m = -n$ is equivalent to paying with the field of view since all the replicas that do not fulfill this condition (crossed terms with $n \neq -m$) will appear at spatial positions of:

$$x_{m,n} = \lambda z_0 v_0(m + n). \quad (2.16)$$

The field of view for the input field distribution is smaller than the expression of (2.16) such that the undesired terms for which m is not equal to $-n$ will not distort the reconstructed image of the output plane.

All this derivation was done before and described in details in refs. [21, 23, 24]. Next it is shown how the usage of polychromatic illumination can remove the drawback of this approach that is related to the payment with the field of view.

Note that the expression (2.15) is for the field distribution. Since the illumination is polychromatic, the intensity for the final outcome of the mathematical derivation is computed and then averaged for the various wavelengths. This is due to the fact that a monochromatic detector averages the readout over the spectral range of the illumination:

$$
\begin{aligned}
|h(x,z=4F)|^2 = \int_{\Delta\lambda} S(\lambda) \sum_m \sum_n \sum_{m'} \sum_{n'} A_m B_n A_{m'}{}^* B_{n'}{}^* \int_{-\infty}^{\infty} \int_{-\infty}^{\infty} \text{rect}\left(\frac{v+mv_0}{\Delta\mu/\lambda 2F}\right) \\
\text{rect}\left(\frac{v'+m'v_0}{\Delta\mu/\lambda 2F}\right) \times \exp\left[2\pi i\left(x(v+v_0(m+n))\right.\right. \\
\left.+ v(z_0\lambda(m+n)v_0) + \frac{z_0\lambda v_0{}^2}{2}(m+n)^2\right)\right] \times \exp\left[-2\pi i\left(x(v'+v_0(m'+n'))\right.\right. \\
\left.\left.+ v'(z_0\lambda(m'+n')v_0) + \frac{z_0\lambda v_0{}^2}{2}(m'+n')^2\right)\right] d\lambda\,dv\,dv',
\end{aligned}
$$

(2.17)

where $/h/^2$ is the intensity impulse response for the spatially incoherent case. $\Delta\lambda$ is the spectral range of the illuminating source (over which the averaging is performed) and $S(\lambda)$ is the spectral distribution of the source. It is assumed as well that this distribution is more or less uniform within the spectral range of $\Delta\lambda$. To obtain the expression for the impulse response, a point source is assumed in the input plane, i.e., its Fourier transform is a constant: $\tilde{u}_0(v) = 1$. In order to compute the output distribution in case that any general distribution is positioned in the input plane, one needs to convolve this impulse response with the intensity of the input object. Let us denote:

$$
\xi = v(z_0(m+n)v_0) + \frac{z_0 v_0{}^2}{2}(m+n)^2 - v'(z_0(m'+n')v_0) - \frac{z_0 v_0{}^2}{2}(m'+n')^2.
$$

Inspecting the obtained result within the spatial spectral range of the synthetic super resolved aperture leads to:

$$
\begin{aligned}
|h(x,z=4F)|^2 = \sum_m \sum_n \sum_{m'} \sum_{n'} A_m B_n A_{m'}{}^* B_{n'}{}^* \int_{-\infty}^{\infty} \int_{-\infty}^{\infty} \text{rect}\left(\frac{v+mv_0}{\Delta\mu/\bar{\lambda} 2F}\right) \\
\text{rect}\left(\frac{v'+m'v_0}{\Delta\mu/\bar{\lambda} 2F}\right) \times \exp[2\pi i x(v+v_0(m+n)-v'-v_0(m'+n'))] \\
\times \int_{\Delta\lambda} S(\lambda)\exp[2\pi i\lambda\xi] d\lambda\,dv\,dv',
\end{aligned}
$$

(2.18)

where $\bar{\lambda}$ is the average wavelength of the illuminating spectral band.

Since the spectral bandwidth of the illumination considered to be wide enough and uniform, it is possible to approximate that:

$$
\int_{\Delta\lambda} S(\lambda)\exp[2\pi i\lambda\xi] d\lambda \approx S(\bar{\lambda}) \int_{\Delta\lambda} \exp[2\pi i\lambda\xi] d\lambda = \delta(\xi).
$$

(2.19)

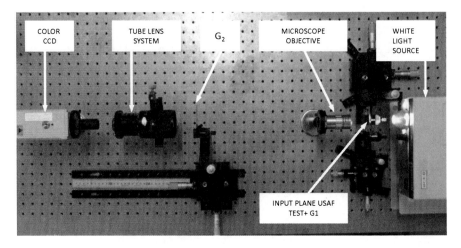

Fig. 2.10 Picture of the experimental setup at the laboratory. The image was taken from: J. García, V. Micó, D. Cojoc and Z. Zalevsky, "Full Field of View Superresolution Imaging based on Two Static Gratings and White Light Illumination," Appl. Opt. 47, 3080–3087 (2008)

Since (2.19) contains a delta function, it is valid only for $\xi = 0$ which is obtained only for the case when the integer indexes fulfill: $m = -n$ and $m = -n$. This is true since only then the phase of the exponent in (2.19) is zero and therefore all the components are added constructively during the integration process over the full range of values of λ. Therefore, the result of (2.15), having the physical meaning of super resolved imaging, may be obtained without limiting the field of view:

$$|h(x, z = 4F)|^2 = \left| \int_{-\infty}^{\infty} \left[\sum_m A_m B_{-m} \text{rect}\left(\frac{v + mv_0}{\Delta\mu / \bar{\lambda} 2F} \right) \right] \exp(2\pi i x v) dv \right|^2. \quad (2.20)$$

The proposed super resolving technique that allows to improve the resolution with two fixed gratings without paying in the field of view still requires the payment in the dynamic range or the signal-to-noise ratio (SNR) in the detector. Nevertheless, availability of detectors with high dynamic range of 12 and more bits turns this drawback into less significant.

2.2.3 Experimental Results

To demonstrate the presented approach, the optical setup shown in Fig. 2.10 is constructed at the laboratory. The experimental setup includes two imaging modules. The magnification of the first imaging system is selected to be 7.5×. The second imaging module magnifies the first image plane into output plane and its magnification can be selected according to our benefit. A long working distance, infinity corrected Mitutoyo microscope lens with 0.14 NA is used as a first imaging system. A photographic objective with variable focus (or magnification) is used as a second imaging system. Notice that, in a similar way as in commercial microscopes, the second imaging system acts as a tube lens. This lens should not have the restriction of having a fixed magnification.

White light illumination is provided by a halogen lamp source and a 3CCD color video camera (SONY Model DXC-950P) captures the final images. The halogen lamp has relatively uniform spectrum in the visible range (it resembles black body radiation) and therefore the assumption for the spectral uniformity as done in the mathematical analysis is valid. The spectrum of the halogen lamp is presented in Fig. 2.11a. These data are taken from the literature. In Fig. 2.11b, the sensitivity response of the three channels (R, G, and B) of the CCD is presented. Those charts are important since what is relevant to the operation principle is not the illuminating spectrum alone but rather its product with the sensitivity of the detector. Figure 2.11c shows the combined result of the camera sensitivity and the spectrum of the illumination by adding the three channels sensitivities each one multiplied by the spectrum irradiance. In order to demonstrate the validity of the assumption for the delta function (2.19), the magnitude of the Fourier transform of the chart of Fig. 2.11d is computed. The display is in dB units. As shown in Fig. 2.11e, the magnitude of the Fourier is indeed nearly a delta function with attenuation of more than ten times the values surrounding the peak of the delta.

Two precision Ronchi ruling slides are used as diffraction gratings in the experiment. The period of both G_1 and G_2 gratings is $p_1 = 600$ lp/mm and $p_1 = 80$ lp/mm, respectively (due to the ratio of magnifications between the two parts of our setup, the second grating could be a low-frequency grating). The period of the first grating is selected depending on the NA of the microscope lens that was used as first imaging system. To achieve a resolution gain factor close to 2, the diffraction angle for a central wavelength of the broadband spectral light used as illumination must be nearly twice the angle defined by the NA of the objective. This means that a period of around 500 lp/mm is suitable for such a resolution improvement. Once the first grating is selected, one can do both: fixing the magnification of the microscope objective and properly selecting the G_2 grating, or the opposite. In our case, a ratio of 7.5 was defined by the periods of both diffraction gratings and this is the magnification that is aimed for the microscope lens.

Since the second imaging setup had a magnification such that the low NA of the imaging lenses did not reduce resolution any more, a true super resolved image was obtained. The experiment was performed for 1-D super resolution and therefore the super resolving factor that was obtained may easily be extracted just by comparing the resulted resolution on both principal axes. Our purpose was to demonstrate the super resolution as well as to show that the result is obtained without paying with the field of view when the white light source is used.

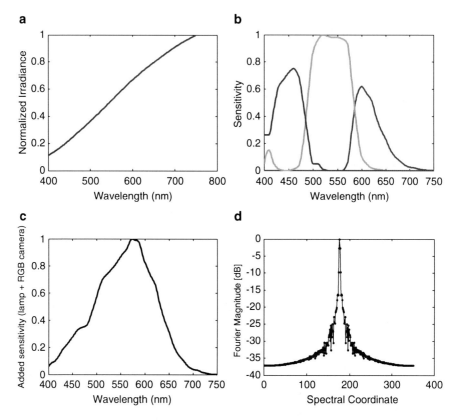

Fig. 2.11 (**a**) The illumination spectrum of halogen lamp. (**b**) Sensitivity response of the three channels (R, G, and B) of the CCD. (**c**) The combined response of the illumination spectrum and the sensitivity of the CCD (addition of the three channel sensitivities each multiplied by the spectral irradiance of the lamp) of (**c**). (**d**) The magnitude of the Fourier transform of the combined chart of (**c**). The image was taken from: J. García, V. Micó, D. Cojoc and Z. Zalevsky, "Full Field of View Superresolution Imaging based on Two Static Gratings and White Light Illumination," Appl. Opt. 47, 3080–3087 (2008)

A negative high-resolution USAF test target was used. Figure 2.12a depicts the full field of view image when the presented approach is used and the magnification of the second imaging system is near to 1. One can see that as the ghost images are wavelength sensitive due to the diffraction orders of the gratings, they are averaged in the background (which means that there is no limitation on the field of view). On the other hand, the proper combination of diffraction orders between both gratings compensates their chromatic dispersion and reinforces the white light super resolved image. In Fig. 2.12b, the classical Bachl and Lukosz monochromatic experiment is shown by simply placing an interference filter (515 nm main wavelength) before the input plane. One may see as the ghost images are not averaged, the final resolution is limited by the distance between the replicated diffraction

Fig. 2.12 Experimental results: (**a**) The full field of view super resolved image obtained using the presented approach, and (**b**) the full field of view image with monochromatic illumination (Bachl and Lukosz approach). (**c**) Cross section of (**a**) for the purpose of computing the reduction in contrast. The image was taken from: J. García, V. Micó, D. Cojoc and Z. Zalevsky, "Full Field of View Superresolution Imaging based on Two Static Gratings and White Light Illumination," Appl. Opt. 47, 3080–3087 (2008)

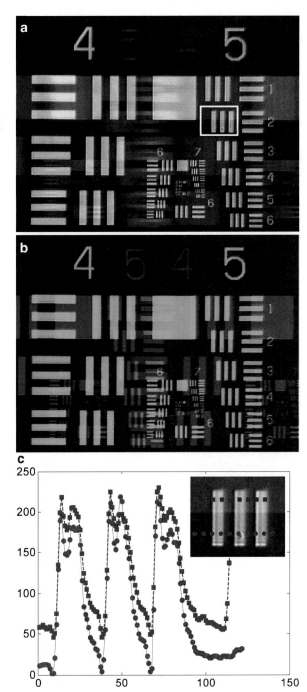

orders. In this case, a reduction in the field of view is needed to allow super resolution over the region of interest.

In Fig. 2.12c, the cross section of the region marked by the square in Fig. 2.12a is shown. The purpose was to compute the reduction in contrast due to the usage of white light illumination. The cross section was computed in two locations (as indicated in the upper right corner of Fig. 2.12c). The red circles indicate the cross section in the lower part of marked region where no replication was generated and thus no reduction in contrast. The blue squares present the cross section in the upper part of the marked region where the various replications (differently positioned due to the usage of the polychromatic illumination) reduced the contrast of the bars. The contrast of the red circles is 0.946 while that of the blue squares is 0.586. This reduction of 39% in contrast is due to the replications. Our computation of contrast was performed according to:

$$C = \frac{I_{max} - I_{min}}{I_{max} + I_{min}}, \tag{2.21}$$

where I_{max} is the maximal value of the intensity and I_{min} is its minimal value.

Note that this super resolution approach as other approaches involving gratings is not energetically efficient. Due to the gratings, only a certain portion of the input energy arrives to the region of interest in the output plane. However, one must distinguish between energetic efficiency and contrast. The reduction in energy may be compensated if the illumination source is strong enough and if the detector has an automatic gain control function that adapts the dynamic range of its sampling (A/D conversion) to the average level of the arriving energy. The contrast reduction cannot be compensated in the hardware since it is related to the SNR and to the number of sampling bits identifying the signal from the background noises.

Theoretically speaking, the reduction in contrast can be estimated as follows: Since the contrast is defined as formulated in (2.21), and due to the replications a D.C background is added to the intensity (to I_{max} as well as to I_{min}), one may obtain the expression for new contrast as:

$$C = \frac{I_{max} - I_{min}}{I_{max} + I_{min} + 2D.C}, \tag{2.22}$$

where the D.C background is exactly the average of the imaged object:

$$D.C = \frac{\int_{\Delta\lambda} S(\lambda) u_0(x - \beta_1\lambda, y - \beta_2\lambda) d\lambda}{\Delta\lambda}, \tag{2.23}$$

where β_1 and β_2 are constants. u_0 is the imaged object. For instance, in the case that resembles our experiment where the object has an average gray level, i.e., D.C of 60 and in spatial region where $I_{max} = 180$ and $I_{min} = 5$, the contrast is reduced to 0.58.

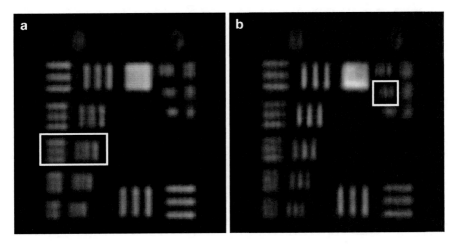

Fig. 2.13 Experimental results showing the high resolution region of interest from Fig. 4a. The reference image is obtained (**a**) without the gratings, and (**b**) with the gratings installed and using the presented super-resolution approach. White squares mark the resolution limit

In Fig. 2.13, one may see the central part of the resolution target where a magnification of close to $7\times$ is chosen for the tube lens system. Indeed, one may see that the resolution of the vertical lines (Group 9, Element 2 corresponding with 575 lp/mm) is much higher than that of the horizontal lines (Group 8, Element 4 corresponding with 362 lp/mm). Therefore, the experiment has demonstrated resolution improvement by a factor of almost 2.

2.3   Super Resolution Using Gray Level Coding

2.3.1   Introduction

The technique presented in the following section provides resolution improvement for both diffractive and geometric limitations [25]. The required constraint is that the object has limited number of gray levels and thus the gray level domain can be used in order to code and decode the additional spatial information. A very interesting application in which the presented super resolving coding may be applied is related to geometrical rather than diffractive super resolution.

2.3.2   Theory

Let us assume that $p(x, y)$ is the blurred point spread function whose blurring is caused due to the combination of the limited aperture of the optics and the area of each pixel in the detection array. The blurred image is sampled by the detection array. "δ_x" and "δ_y" are denoted as the sampling pitch in the horizontal and vertical axes, respectively. Δx and Δy symbolize the horizontal and the vertical dimensions of the pixels in the detection array, respectively. Thus, the sampled image equals to:

$$I_o(x,y) = \int_{-\Delta x/2}^{\Delta x/2} \int_{-\Delta y/2}^{\Delta y/2} I_{in}(x',y')C(x',y')p(x-x',y-y')dx'dy' \sum_n \sum_m \delta(x-n\delta_x, y-m\delta_y),$$

(2.24)

where $C(x,y)$ is the gray level coding mask. The last equation equals to:

$$I_o(n\delta_x, m\delta_y) = \sum_n \sum_m \left[\int_{n\delta_x-\theta_x/2}^{n\delta_x+\theta_x/2} \int_{m\delta_y-\theta_y/2}^{m\delta_y+\theta_y/2} I_{in}(x',y')C(x',y')p(n\delta_x - x', m\delta_y - y')dx'dy' \right]$$
$$\times \delta(x-n\delta_x, y-m\delta_y),$$

(2.25)

where θ_x and θ_y are the horizontal and the vertical dimensions of the blurring function, $p(x, y)$, respectively. For the simplicity of explanation, it is assumed that the input object I_{in} is a binary function, having resolution coinciding with the detectors sampling grid δ_x and δ_y:

$$I_{in}(k_1\delta_x, k_2\delta_y) = \{0,1\}.$$

(2.26)

To simplify further, it is assumed that $p(x, y)$ is a rect function:

$$p(x,y) = \text{rect}\left(\frac{x}{\theta_x}, \frac{y}{\theta_y}\right),$$

(2.27)

where $\theta_x = N\delta_x$ and $\theta_y = M\delta_y$, N and M are integer numbers.
The gray level coding mask is chosen such that:

$$C(x,y) = \left[\sum_{k_1=n-N/2}^{n+N/2-1} \sum_{k_2=m-M/2}^{m+M/2-1} 2^{k_1+N/2-n} \cdot 2^{k_2+M/2-m} \text{rect}\left(\frac{x-k_1\delta_x}{\delta_x}, \frac{y-k_2\delta_y}{\delta_y}\right) \right]$$
$$* \delta(x-nN\delta_x, y-mM\delta_y).$$

(2.28)

Thus, the output intensity equals to:

$$I_o(n\delta_x, m\delta_y) = \sum_{k_1=n-N/2}^{n+N/2-1} \sum_{k_2=m-M/2}^{m+M/2-1} I_{in}(k_1\delta_x, k_2\delta_y) \cdot 2^{k_1+N/2-n} \cdot 2^{k_2+M/2-m}. \quad (2.29)$$

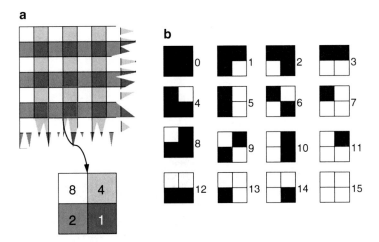

Fig. 2.14 (a) The gray level coding mask. (b) The look up table relating sensed gray level and the spatial structure of the original object. The image was taken from: Z. Zalevsky, P. García-Martínez and J. García, "Superresolution Using Gray Level Coding," Opt. Exp. 14, 5178–5182 (2006)

The meaning of the last equations is that the coding mask is chosen such that it is actually the binary base. Therefore, after blur is applied, the gray level of the blurred pixel equals to a different gray level. For instance, let us assume that the super resolution is a factor of 2 in each dimension, i.e., $N = 2$ and $M = 2$, then the coding mask is a periodic structure with super pixels constructed out of blocks with a gray level of 1, 2, 4, and 8 as shown in Fig. 2.14a. After blurring and assuming that I_{in} is a binary object, the resulted gray level will indicate the spatial structure of I_{in} prior to blurring. In Fig. 2.14b, a look up table is shown which connects the spatial structure of I_{in} in the super pixel prior to blurring and the resulted gray level, when the structure is multiplied by the gray level coding mask and integrated over a super pixel (2.29).

2.3.3   Experiment

The experimental setup is depicted in Fig. 2.15. Spatial light modulator (SLM) was attached to a binary object. The gray coding mask was displayed on the SLM. The light passing through the object and the mask was imaged on the top of a camera. For the purpose of demonstration, a detector binning was used to simulate a low-resolution device. This permits to record also a high-resolution version for comparison.

A binning of 1 by 5 was performed in the camera. The coding mask displays gray values of 1, 2, 4, 8, and 16. The imaged object seen by the camera without applying the binning (high resolution) is presented in Fig. 2.16.

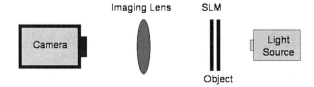

Fig. 2.15 The experimental setup. The image was taken from: Z. Zalevsky, P. García-Martínez and J. García, "Superresolution Using Gray Level Coding," Opt. Exp. 14, 5178–5182 (2006)

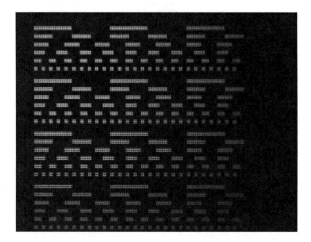

Fig. 2.16 The high-resolution image. The image was taken from: Z. Zalevsky, P. García-Martínez and J. García, "Superresolution Using Gray Level Coding," Opt. Exp. 14, 5178–5182 (2006)

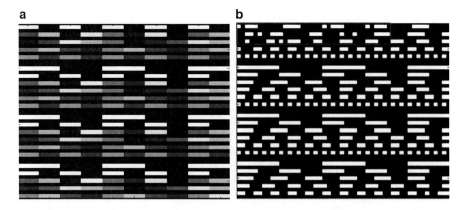

Fig. 2.17 (**a**) The experimentally grabbed image after binning and before decoding. (**b**) The experimentally reconstructed image after the decoding of gray levels. The image was taken from: Z. Zalevsky, P. García-Martínez and J. García, "Superresolution Using Gray Level Coding," Opt. Exp. 14, 5178–5182 (2006)

Figure 2.17a presents the image seen by the camera after applying the binning. One may see that most of the spatial content of the object is lost due to the binning of low-pass effect. Figure 2.17b displays the experimentally reconstructed image after decoding the gray levels. One may see that except for few reconstruction errors, which are outlined by red box, the original spatial resolution of the object was reconstructed.

Note that the suggested gray coding that translates the captured gray level into resolution of the original image is a very simple code. This code is not immune to errors. This is something which is very nonrecommendable since small error in the gray level may change completely the decoding pattern (see, for instance, the variations of the decoded patters vs. the gray level in Fig. 2.14b). However, it is very simple to use an optical code that is much more immune to gray level errors. One example can be the Gray codes [26]. In those codes, the change between two adjacent codes has variation of only 1 bit and thus errors in the gray level will cause minimal spatial distortion (distortion of only 1 bit).

References

1. Johnson, R.B., Feng, C.: Mechanically compensated zoom lenses with a single moving element. Appl. Opt. **31**, 2274–2280 (1992)
2. Tam, E.C.: Smart electro optical zoom lens. Opt. Lett. **17**, 369–371 (1992)
3. Tsuchida, H., Aoki, N., Hyakumura, K., Yamamoto, K.: Design of zoom lens systems that use gradient-index materials. Appl. Opt. **31**, 2279–2286 (1992)
4. Pegis, R.J., Peck, W.G.: First-order design theory for linearly compensated zoom systems. J. Opt. Soc. Am. **52**, 905–911 (1962)
5. Wooters, G., Silvertooth, E.W.: Optically compensated zoom lens. J. Opt. Soc. Am. **55**, 347–355 (1965)
6. ChunKan, T.: Design of zoom system by the varifocal differential equation. I. Appl. Opt. **31**, 2265–2273 (1992)
7. Ito, Y.: Complicated pin-and-slot mechanism for a zoom lens. Appl. Opt. **18**, 750–758 (1979)
8. Shafer, D.R.: Zoom null lens. Appl. Opt. **18**, 3863–3870 (1979)
9. Tanaka, K.: Paraxial analysis of mechanically compensated zoom lenses. 1: Four-component type. Appl. Opt. **21**, 2174–2181 (1982)
10. Zhang, D.Y., Justis, N., Lo, Y.H.: Integrated fluidic adaptive zoom lens. Opt. Lett. **29**, 2855–2857 (2004)
11. Walter, A.: Zoom lens and computer algebra. J. Opt. Soc. Am. A **16**, 198–204 (1999)
12. Akram, M.N., Asghar, M.H.: Step-zoom dual-field-of -view infrared telescope. Appl. Opt. **42**, 2312–2316 (2003)
13. Walther, A.: Angle eikonals for a perfect zoom system. J. Opt. Soc. Am. A **18**, 1968–1971 (2001)
14. Zalevsky, Z., Zlotnik, A.: Single snap-shot double field optical zoom. Opt. Exp. **13**, 9858–9868 (2005)
15. Solomon, J., Zalevsky, Z., Mendlovic, D.: Geometrical super resolution by code division multiplexing. Appl. Opt. **44**, 32–40 (2005)
16. Dammann, H., Klotz, E.: Coherent optical generation and inspection of two-dimensional periodic structures. Opt. Acta **24**, 505–515 (1977)
17. Lukosz, W.: Optical systems with resolving powers exceeding the classical limits. J. Opt. Soc. Am. **56**, 1463–1472 (1966)

18. Bachl, A., Lukosz, W.: Experiments on superresolution imaging of a reduced object field. J. Opt. Soc. Am. **57**, 163–169 (1967)
19. Sabo, E., Zalevsky, Z., Mendlovic, D., Konforti, N., Kiryuschev, I.: Super resolution optical system using two fixed generalized Dammann gratings. Appl. Opt. **39**, 5318–5325 (2000)
20. Zalevsky, Z., Mendlovic, D., Lohmann, A.W.: Super resolution optical systems using fixed gratings. Opt. Commun. **163**, 79–85 (1999)
21. Sabo, E., Zalevsky, Z., Mendlovic, D., Konforti, N., Kiryuschev, I.: Super resolution optical system using three fixed generalized gratings: experimental results. J. Opt. Soc. Am. A **18**, 514–520 (2001)
22. García, J., Micó, V., Cojoc, D., Zalevsky, Z.: Full field of view superresolution imaging based on two static gratings and white light illumination. Appl. Opt. **47**, 3080–3087 (2008)
23. Zalevsky, Z., Mendlovic, D., Lohmann A.W.: Progress in optics. In: Wolf, E. (ed.) Optical System with Improved Resolving Power, vol. XL, Chap. 4. North Holland, Amsterdam, (1999)
24. Zalevsky, Z., Mendlovic, D.: Optical Super Resolution. Springer, New York (2003)
25. Zalevsky, Z., García-Martínez, P., García, J.: Superresolution using gray level coding. Opt. Exp. **14**, 5178–5182 (2006)
26. Gray, F.: Pulse code communication. US Patent 2,632,058, 17 Mar 1953

Chapter 3
Techniques Utilizing Diffractive Masks Having Structures with a Period Limited Randomness

Alex Zlotnik, Zeev Zalevsky, David Mendlovic,
Jonathan Solomon, and Bahram Javidi

3.1 Geometrical Super Resolution Using Code Division Multiplexing

3.1.1 Introduction

In many high resolving optical systems, resolution is limited not by the optics but by the sensor's nonzero pixel size. As a result, overall resolution is decreased. Here, a novel approach for enhancing resolution beyond the limit set by the sensor's pixels is proposed. This method does not involve additional mechanical elements, such as those used for microscan. In this scheme, neither the sensor nor additional elements are moved. The geometrical super resolving procedure is based on code division multiplexing access (CDMA) approach with all of its inherent benefits, such as relative noise immunity to single tone interference. A setup is proposed for coherent and incoherent illumination, with slight modifications for the latter. A theoretical analysis of the setup is presented and later compared with empirical results.

This scheme is shown to enhance a one-dimensional image resolution with the use of only a simple mask which doubled image resolution. This method can easily be expanded to two-dimensional images and resolution enhancement factors are greater than two.

Z. Zalevsky (✉)
School of Engineering, Bar-Ilan University, Ramat-Gan, Israel
e-mail: zalevsz@macs.biu.ac.il

Z. Zalevsky (ed.), *Super-Resolved Imaging: Geometrical and Diffraction Approaches*, 39
SpringerBriefs in Physics, DOI 10.1007/978-1-4614-0833-8_3,
© Springer Science+Business Media, LLC 2011

Fig. 3.1 CDMA signals.
(*Top*) The waveform of the
data stream; (*middle*) the
chipping waveform; (*bottom*)
the waveform product

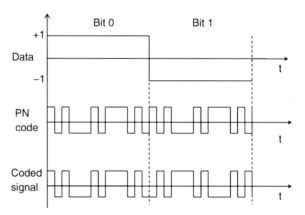

3.1.2   Theoretical Analysis

As the sensor samples the image, aliasing is produced and the image is distorted. This is due to the fact that the spectral bandwidth of the physical image is relatively large. Data corruption occurs since it is not possible to discriminate the different parts of the spectrum in overlapping regions. This is similar to a common problem in communication: the need to transmit several messages on a common resource, such as an electronic wire. The solution for this is multiplexing. Each message is coded in such a way that it can be retrieved later. Common multiplexing schemes involve frequency multiplexing, time division multiplexing, and CDMA. The latter offers varied advantages, which is further elaborated in Sect. 3.1.2.1

To avoid data loss in regions of the spectrum that will overlap, potentially overlapping sections of the spectrum are coded with different spectral masks. These masks will be orthogonal as is required in CDMA. Coding needs to be in the Fourier plane, as replicas (due to sampling) are created in the Fourier plane.

Assuming that different parts of the spectrum are coded correctly, data corruption can be prevented and the received image can be enhanced [1].

3.1.2.1   Code Division Multiplexing

Code division multiplexing access (CDMA) technology focuses primarily on the "direct sequence" method of spread spectrum [2]. Direct sequence is spread spectrum technology in which the bandwidth of a signal is enlarged by artificially increasing the bit data rate by breaking each bit into a number of subbits called "chips."

The signal is divided into smaller bits by multiplying it with a pseudonoise (PN) code. A simple multiplication of each bit of the original modulated signal by this high data rate PN code yields the division of the signal into smaller bits (which increases its bandwidth). Increasing the number of "chips" expands the bandwidth proportionally. This is demonstrated in Fig. 3.1.

Fig. 3.2 Optical setup for
CDMA super resolution

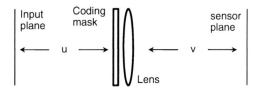

Let us now briefly describe the basic operation of the transmitter/receiver for the spread spectrum technique. Assume that there are two transmitters with two different messages to be transmitted. The messages are first modulated. After the modulator, each signal is multiplied by its own unique pseudonoise code and is transmitted. Since various signals might be simultaneously transmitted from different transmitters, these transmissions are represented by simply adding their spectra. At the receiver end, the incoming signal is the spread spectrum signal. In order to extract a single message, one must multiply the incoming signal by the corresponding PN code. Multiplying a given PN code by itself will produce unity. Therefore, multiplying the signal by the PN code eliminates the spread spectrum effects for that particular message. This is of course valid for orthogonal codes with perfect synchronization and no noise. The resulting signal is then passed through a band pass filter (BPF) centered at the carrier frequency. This operation selects only the desired signal while rejecting all surrounding frequencies due to other messages in the spread spectrum.

This scheme is used extensively in communication to provide multiuser access where each user uses a unique PN code. This method provides a rather significant single tone interference immunity, which is important in imaging, and a trivial optical implementation (a simple binary mask).

3.1.2.2   Optical Setup

For achieving superresolution in the presented approach, the following steps are required:

1. Fourier transform of the object
2. Multiplication by a CDMA coding mask
3. Inverse Fourier transform
4. Sampling the output
5. Retrieval of the object in full resolution

The optical setup can be very simple, as the only additional requirement apart from a simple imaging system (containing only one lens and a sensor) is to multiply the image with a mask in the Fourier plane. This can be achieved by placing the mask against the lens [3]; special care must be taken for verifying the different dimension scales needed. The optical setup is presented in Fig. 3.2.

Note that since the lens aperture is not the cause for the resolution reduction, the coding mask may be attached to the lens knowing that it will perform the required orthogonal coding of the various spectral regions, later on to be aliased due to the sampling of the sensor.

3.1.2.3 Mathematical Analysis

Let us examine a simple situation, with the resolution enhanced by a factor of 3. Only a one-dimensional calculation is carried out, carrying out the proof for two-dimensions is direct. An ideal sensor is assumed, in which the pixels are indefinitely small and are placed at a distance of δx from one another (finite size pixels are addressed later). Therefore, the sensor is modeled as an infinite train of impulses:

$$\text{Sens}(x) = \sum_{n=-\infty}^{\infty} \delta(x - n\,\delta x). \tag{3.1}$$

The coding mask is divided into three subfunctions as follows:

$$\tilde{G}(v) = \sum_{n=-1}^{1} g_n(v - n\Delta v), \tag{3.2}$$

where $\Delta v = 1/\,\delta x$, g_i has the following properties:

$$\begin{cases} g_i(v)g_j(v - \Delta v) = 0, & i \neq j, \\ g_i(v) = 0, & i = -1, 0, 1, \quad \forall\, v \notin \left\{ -\dfrac{3\Delta v}{2}, \dfrac{3\Delta v}{2} \right\}, \\ g_i(v) \geq 0, & \forall v. \end{cases} \tag{3.3}$$

These properties promise orthogonality of the coding masks. This is shown graphically in Fig. 3.3. The signals in the Fourier plane (the aperture plane) have the coordinate "v," and the subscript "$\sim$." Signals in the sensor plane have the coordinate "x" and have no subscript.

Notice that the coding masks have been chosen to be nonnegative. The masks are composed of pixels of the size of $\Delta\eta$. Each pixel is divided into chips, each have the size of ΔW. The consideration for this are presented later on. The coding mask is multiplied in the Fourier plane with the spectrum of the input, I (which represents field distribution). Therefore, the output in the Fourier plane is:

$$\tilde{O}(v) = \tilde{I}(v)\text{Sen}\tilde{s}(v). \tag{3.4}$$

The sensor samples this output; therefore, the sampled output $S(x)$ is:

$$S(x) = O(x)\text{Sens}(x). \tag{3.5}$$

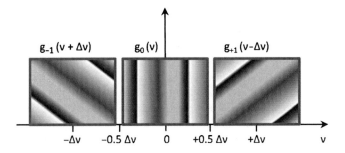

Fig. 3.3 Coding mask

Expressing equation (3.5) in the Fourier plane, and using (3.1), (3.2), and (3.4) yields:

$$\tilde{S}(v) = \tilde{O}(v) * \text{Sen}\tilde{s}(v) = \left[\tilde{I}(v) \sum_{n=-1}^{1} g_n(v - n\Delta v) \right] * \left[\sum_{n=-\infty}^{\infty} \delta\left(v - n\frac{1}{\Delta x} \right) \right], \quad (3.6)$$

where $*$ denotes convolution operation. Since $\Delta v = 1/\Delta x$, the last expression is simplified to:

$$\tilde{S}(v) = \left[\tilde{I}(v) \sum_{n=-1}^{1} g_n(v - n\Delta v) \right] * \left[\sum_{n=-\infty}^{\infty} \delta(v - n\Delta v) \right],$$

$$= \sum_{n=-\infty}^{\infty} \tilde{I}(v - n\Delta v) \left[\sum_{k=-1}^{1} g_n(v - (n + k)\Delta v) \right]. \quad (3.7)$$

Image retrieval is simply achieved by Fourier transforming the grabbed output and multiplying it with the original coding mask:

$$\tilde{R}(v) = \tilde{S}(v)\tilde{G}(v)$$

$$= \left\{ \sum_{n=-\infty}^{\infty} \tilde{I}(v - n\Delta v) \left[\sum_{k=-1}^{1} g_n(v - (n + k)\Delta v) \right] \right\} \left[\sum_{m=-1}^{1} g_m(v - m\Delta v) \right],$$

$$= \sum_{n=-\infty}^{\infty} \tilde{I}(v - n\Delta v) g_n(v - n\Delta v),$$

$$= \tilde{I}(v)\tilde{G}(v). \quad (3.8)$$

Choosing only the output field quantities inside the nonzero chip of each pixel for the given coding mask, a down sampled output is generated which is the desired output, $I(v)$.

The conclusion is that modulating the input's spectrum correctly prevents data corruption due to aliasing. The input of the optical imaging system is assumed to be only real and nonnegative. It is convolved with the Fourier transform of the coding mask which is also real since it is originally symmetrical in the Fourier plane. The coding mask should be chosen in such a way that there are no negative values in the image plane. This is further elaborated in Sects. 3.1.2.5 and 3.1.2.6 dealing with different types of illumination. In a real sensor, the pixels have finite dimensions; this will affect the output spectrum, since it results with multiplying the spectrum with a $\sin c$ function, as mentioned above. This will have no effect on spectrum orthogonality; it will only introduce a need for recalibrating the sampled spectrum, but does not introduce any difficulties.

In this proof, it was assumed that the sensor performs ideal sampling, yet physically it measures only absolute values of the sampled signal. If the coding mask is chosen correctly, then this limitation can be overcome. This is elaborated further on.

3.1.2.4 The Effect of Noise

The super resolution scheme presented here offers no significant advantage or disadvantage upon any other method before the signal is coded. Therefore, if in the original image a noise is present, then its reconstruction will have the same noise. The scheme will provide an advantage on noise accumulated after the image is coded, such as during sampling by the sensor. Let us assume that the sensor is scratched or dirty. This presents a very narrow interference multiplied by the input function. The interference function is represented by $n(x)$. The signal measured now by the sensor, $S'(x)$, is equal to:

$$S'(x) = S(x)n(x). \tag{3.9}$$

The retrieved output, $\tilde{R}'(v)$, equals:

$$\tilde{R}'(v) = \left[\tilde{S}(v) * \tilde{n}(v)\right]\tilde{G}(v) = \left[\tilde{S}(v)\tilde{G}(v)\right] * \left[\tilde{n}(v)\tilde{G}(v)\right]. \tag{3.10}$$

According to (3.8), this can be expressed as:

$$\tilde{R}'(v) = \left[\tilde{I}(v)\tilde{G}(v)\right] * \left[\tilde{n}(v)\tilde{G}(v)\right]. \tag{3.11}$$

In the space domain, this equals to:

$$R'(x) = [I(x) * G(x)][n(x) * G(x)]. \tag{3.12}$$

Since the function G is assumed to be pseudowhite noise in the object plane and since the noise function, $n(x)$, is very narrow:

$$n(x) * G(x) \approx \text{const} + \varepsilon G(x - x_0), \tag{3.13}$$

where const $\gg \varepsilon$. Using (3.11), the retrieved image is expressed as:

$$R'(v) = \text{const}\, I(v)\tilde{G}(v) + \varepsilon \exp(-2\pi i x_0)\left[I(v)\tilde{G}(v)\right] * \tilde{G}(v). \tag{3.14}$$

Since the noise factor is constant and relatively negligible, one can express the above equation as:

$$R'(v) \approx \text{const}\, \tilde{I}(v)\tilde{G}(v). \tag{3.15}$$

From here, retrieval is identical to what was presented for noninterfered signal. The advantage that CDMA coding provides for overcoming very space limited interference was intuitively proven. This is identical to communication CDMA scheme's improved resistance to single tone interference (more detail in ref. [4])

3.1.2.5 Coherent Illumination

When coherent illumination is used, the coherent transfer function, $\text{CT}\tilde{F}(v)$, represents the resolution limitations due to diffraction and additional elements in the system. The $\text{CT}\tilde{F}(v)$ equals to the aperture of the imaging lens multiplied by the coding mask, $\tilde{G}(v)$, which is attached to it. The output image in the sensor plane, O_i, is expressed as:

$$O_i(x) = \text{CTF}(x) * g_g(x). \tag{3.16}$$

where g_g indicates the object obtained at the sensor plane in an ideal system (without any other aberrations). The sensor samples only magnitudes of the image. Therefore, the sampled image, I_i, equals to:

$$I_i = \left|\text{CTF}(x) * g_g(x)\right|^2. \tag{3.17}$$

In order not to have information loss, system output, O_i, should be real and nonnegative; this will enable retrieval of output directly from I_i (which is actually the squared absolute of O_i). In order to ensure direct retrieval of the object, g_g in full resolution, both the coding mask and the image must be real and nonnegative. Since real images have no negative values, therefore it is left to deal only with the coding mask. The coding mask must have a Fourier transform that is real and nonnegative. Furthermore, the coding mask must fulfill (3.3). Below an example for a suitable coding mask is presented:

Let us choose a coding mask composed of an infinite train of delta-functions in the space domain, each impulse is spaced by $\Delta\eta$ from adjacent impulses. This signal has a positive spectrum. Now let us convolve this with a Gaussian, with the width of a CDMA chip, ΔW. This will be multiplied by a rect function, setting the size of the mask to finite size, $\Delta BW \sqrt{b^2 - 4ac}$. ΔBW corresponds to the aperture of the imaging lens. Let us now calculate the spectrum of such a mask:

$$\widetilde{CTF}(v) = \{ \text{ Infinite pulse train} * \text{Gaussian} \}\text{rect function},$$

$$\Downarrow$$

$$\widetilde{CTF}(v) = \left\{ \left[\sum_{n=-\infty}^{\infty} \delta(v - n\Delta\eta) \right] * \exp\left(-\frac{v^2}{2\Delta W} \right) \right\} \text{rect}\left(\frac{v}{\Delta BW} \right),$$

$$\Downarrow$$

$$CTF(x) = \Delta W \sqrt{2\pi} \left\{ \left[\sum_{n=-\infty}^{\infty} \delta\left(x - n\frac{2\pi}{\Delta\eta} \right) \right] \exp\left(-\frac{x^2 \Delta W^2}{2} \right) \right\} * \sin c\left(\frac{x\Delta BW}{2} \right).$$

$$(3.18)$$

Assuming that $\Delta BW \gg \Delta W$, a real and nonnegative spectrum was received as desired. One can easily show that this mask also satisfies the conditions for CDMA coding, if the following equation is realized:

$$\Delta v = k\Delta\eta - \frac{\Delta W}{2}, \quad k \in N. \tag{3.19}$$

This is demonstrated graphically in Fig. 3.4. The top figure presents an input object spatial spectrum with a spatial bandwidth of $2\Delta v$. An example of coding mask is shown in Fig. 3.4b. Figure 3.4c demonstrates schematically the effect of coding mask on object spatial spectrum. The bottom figure, Fig. 3.4d, demonstrates the effect of sampling by the sensor. Notice that data retrieval is possible if the orthogonality was retained. This example is suitable for the coding mask.

3.1.2.6 Incoherent Illumination

In incoherent illumination, there is a little variation. The intensity distribution sampled by the sensor is expressed as:

$$\Im(I_i) = \widetilde{OTF}(v)\left(\tilde{U}_g(v) \otimes \tilde{U}_g(v)\right), \tag{3.20}$$

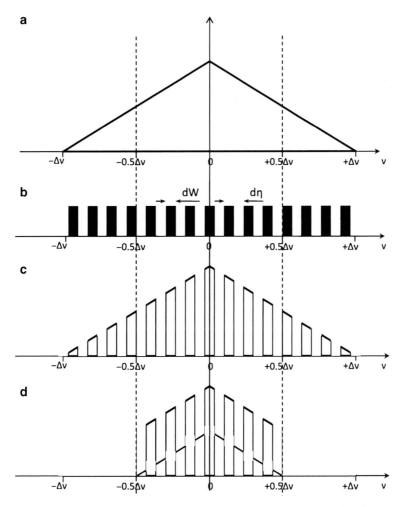

Fig. 3.4 (**a**) Object spatial spectrum; (**b**) example for coding mask, with *black rectangles* schematically illustrating Gaussian chips; (**c**) object spectrum multiplied by coding mask; (**d**) image spectral distribution after down sampling by sensor. Notice no overlapping occurred

where $\text{OT}\tilde{\text{F}}(v)$ is the incoherent optical transfer function. $\otimes$ symbolizes correlation operation. Goodman shows a more detailed derivation of these relations in ref. [2].

The usage of incoherent illumination affects the output of the system. Not the coding mask itself should be orthogonal, but rather its autocorrelation. Let us express the incoherent optical transfer function as a product of autocorrelation of the coherent transfer function:

$$\text{OT}\tilde{\text{F}}(v) = \left(\text{CT}\tilde{\text{F}}(v) \otimes \text{CT}\tilde{\text{F}}(v)\right). \tag{3.21}$$

This can be expressed in the sensor pane as:

$$\text{OTF}(x) = \text{CTF}(x)\overline{\text{CTF}(x)}, \tag{3.22}$$

where $\widetilde{\text{OTF}}(v)$ is the autocorrelation of the previous function $\widetilde{\text{CTF}}(v)$, and its Fourier transform satisfies several conditions:

$$\begin{cases} \widetilde{\text{OTF}}(v=0) \geq 0 \\ \widetilde{\text{OTF}}(v) = \widetilde{\text{OTF}}(-v) \\ \widetilde{\text{OTF}}(v=0) \geq \left|\widetilde{\text{OTF}}(v)\right| \end{cases} \begin{cases} \text{OTF}(x) \text{ is Real,} \\ \text{OTF}(x) \geq 0, \\ \text{OTF}(x) = \text{OTF}(-x). \end{cases} \tag{3.23}$$

Furthermore, the $\widetilde{\text{OTF}}(v)$ has to satisfy all of the previous demands for the CDMA coding, i.e., those presented in (3.3). Again one faces the challenge of producing a coding mask, which has a real and nonnegative spectrum. This problem has been addressed above for the mask implemented for incoherent illumination. The final mask $\widetilde{\text{CTF}}(v)$ will be easily derived since according to Sect. 3.1.2.3 $\widetilde{\text{OTF}}(v)$ that satisfies all of the demands could be constructed. The mask that will be placed in the setup will be $\widetilde{\text{CTF}}(v)$. Since it is real and nonnegative, the derivation is direct from (3.21):

$$\text{OTF}(x) = [\text{CTF}(x)]^2$$
$$\Downarrow$$
$$\text{CTF}(x) = \sqrt{\text{OTF}(x)}. \tag{3.24}$$

3.1.2.7 The Price of Super Resolution

The trade-offs of this system are presented below:

(i) Loss of field of view – Since the spectrum of the image is multiplied with a high-resolution coding mask, the image is spread in the image plane. Since this spread image must not overlap with other images, the field of view of the system must be limited, as not fulfilling this would distort the original input. Therefore, the field of view must be limited, by a factor, which is identical to the expected resolution improvement, i.e., to have a resolution enhancement of three, the original image must not cover more than one-third of the sensor.

(ii) Loss of energy – Possible energy loss occurs only due to the fact that a coding mask is placed over the input lens.

3.1.2.8 The Two-Dimensional Case

Extension of this method to two-dimensional images is direct. Construction of the coding mask should be orthogonal in both axes. Such a mask placed on the lens of an identical optical setup will produce a resolution enhancement of M on the first

Table 3.1 Physical value used for simulation

Attribute	Value	Units
Number of sensor pixels	640	–
Pixel spacing, Δx	9.6	μm
Pixel size, Δd	6.72	μm

axis and N on the second axis (depending on mask attributes on each axis). This will cause a loss of the total field of view of $M \times N$. Image retrieval will also be identical, as a two-dimensional Fourier transform can be conducted separately for each axis.

3.1.3   Computer Simulations

In order to verify the method presented in previous sections, a MATLAB simulation was constructed. This simulation assumed real dimensions of sensor pixel size and pixel spacing, which are shown in Table 3.1. For simplicity, the simulation is one-dimensional. Furthermore, the simulation dealt with a coherent system. An input image of a cosine grating was chosen, shown in Fig. 3.5. This cosine was at a frequency in which sensor sampling will cause obvious aliasing. A coding mask was constructed as described above, as shown in Fig. 3.6.

In the simulation, the input was Fourier transformed, multiplied with the coding mask and inverse Fourier transformed again. This simulated the coding mask attached to the imaging lens. The magnitude of the output was sampled according to attributes of the sensor. Signal retrieval was conducted as follows: sampled data was inverse Fourier transformed, multiplied by the coding mask, and Fourier transformed to produce the desired output.

The simulated output of the system is presented in Fig. 3.7: The top image presents the ideal output of the system (which is actually the original object that is to be imaged by the system); the middle image presents the output without applying the super resolution method. One can see a complete loss of image resolution, and obvious aliasing, as the image frequency appears much lower than that of the original object. The bottom image shows the input after reconstruction using CDMA super resolution; one can see that the image was satisfactory reconstructed.

To further elaborate the functionality of the system, the Fourier domain should be introduced. In the top image in Fig. 3.8, the spectrum of the original cosine grid input is shown. The rectangle illustrates the allowed bandwidth due to sampling by the sensor. In the middle figure, the spectrum of the sampled output is presented. Many artifacts have been added to the spectrum due to multiplication with the coding mask. Notice in the lower figure that after multiplying the spectrum by the coding mask, all aliased frequencies are removed, leaving the original input, i.e., perfect reconstruction.

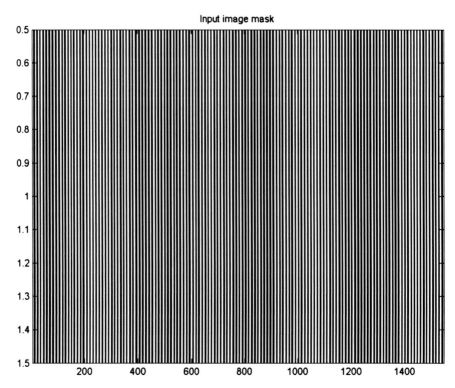

Fig. 3.5 Cosine grid used as simulation input. The image was taken from: J. Solomon, Z. Zalevsky and D. Mendlovic, "Geometrical Super Resolution by Code Division Multiplexing," Appl. Opt., 44, 32–40 (2005)

3.1.4 Experimental Results

In order to verify the method presented, an experimental setup was constructed. The setup consists of a simple imaging system, with only a single lens, an aperture place in the focal plane (used to limit system bandwidth), and a sensor. The coding mask was attached to the lens. A sensor, with attributes as illustrated in Table 3.1, sampled the output. The same cosine input grid and coding mask, as were presented in the simulation, were used for the experiment.

The sampled output is shown in the top of Fig. 3.9. The expansion of the image due to the usage of the coding mask is apparent when comparing with the output width without coding mask, as appears in the lower image on the same figure. Figure 3.10 allows appreciation of performance: The top image shows ideal output. The middle image shows the sampled output without the coding mask (this is identical to Fig. 3.7 (bottom) only the scale is different). One notices the obvious aliasing by the appearance of lower frequencies. Finally, the lower graph shows the retrieved output using CDMA super resolution. This CDMA method produced

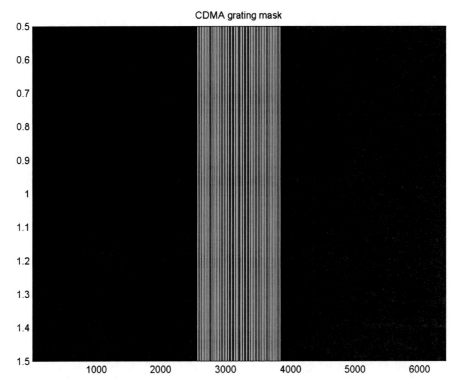

Fig. 3.6 Coding mask used for simulation. The image was taken from: J. Solomon, Z. Zalevsky and D. Mendlovic, "Geometrical Super Resolution by Code Division Multiplexing," Appl. Opt., 44, 32–40 (2005)

satisfactory results for the reconstruction. Notice that a slight degradation of the lower frequencies appears in the retrieved image. Theoretically, this should not be appeared, it is probably a result of a small alignment error between the sampled image and the coding mask.

3.2   Diffraction Super Resolution Using Code Division Multiplexing

3.2.1   Introduction

In this section, CDMA is used to overcome the resolving power of an optical imaging system, and not by correcting the data obtained on the detector plane as described in Sect. 3.1. To enable such multiplexing, a unique setup that creates an incoherent cosine transform of the image is used.

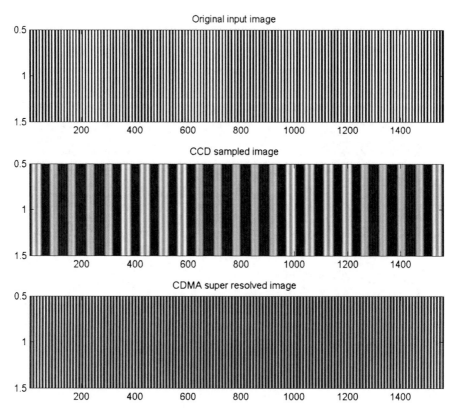

Fig. 3.7 (*Top*) Cosine grid, used as input; (*middle*) cosine grid after sampling by sensor – notice aliasing; (*bottom*) signal retrieved using CDMA super resolution. The image was taken from: J. Solomon, Z. Zalevsky and D. Mendlovic, "Geometrical Super Resolution by Code Division Multiplexing," Appl. Opt., 44, 32–40 (2005)

3.2.2 Theoretical Analysis

3.2.2.1 Example of Analysis of Super Resolution in Wigner Space

Super Resolution can be achieved by placing several gratings at specific locations within the imaging setup. Lukosz [5] has presented a setup based on two static Ronchi gratings that achieve this. Figure 3.11 shows the optical setup used for achieving the super resolution. One may see from the optical setup that the system contains two gratings placed one after the input plane and one after the output plane. These distances and the gratings frequency must be chosen carefully according to conditions described in ref. [5].

The various optical and mathematical steps are:

- $u_0(x, 0) \rightarrow u_0(x, z_0^-)$ free space propagation
- $u_0(x, z_0^-) \rightarrow u_0(x, z_0^+)$ passing through grating A

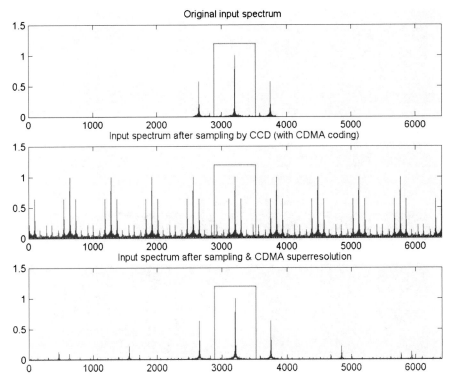

Fig. 3.8 (*Top*) Cosine grid, used as input, spectrum; (*middle*) cosine grid after sampling by sensor; (*bottom*) signal retrieved using CDMA superresolution. The image was taken from: J. Solomon, Z. Zalevsky and D. Mendlovic, "Geometrical Super Resolution by Code Division Multiplexing," Appl. Opt., 44, 32–40 (2005)

- $u_0(x, z_0{}^+) \rightarrow u_0(x, 0)$ virtual backwards propagation
- $u_0(x, 0) \rightarrow u_0(\mu, 2F-)$ optical Fourier transform
- $u_0(\mu, 2F-) \rightarrow u_0(\mu, 2F-) \, \text{rect}(\mu/\Delta\mu) = u_0(\mu, 2F+)$ passing through an aperture
- $u_0(\mu, 2F+) \rightarrow u_0(x, 4F)$ optical Fourier transform
- $u_0(x, 4F) \rightarrow u_0(x, (4F + z_1)^-)$ free space propagation
- $u_0(x, (4F + z_1)^-) \rightarrow u_0(x, (4F + z_1)^+)$ passing through grating B
- $u_0(x, (4F + z_1)^+) \rightarrow u_0(x, 4F)$ virtual backwards propagation

Now let us examine this setup in the Wigner space. This is presented in Fig. 3.12. In Fig. 3.12a, the SW product of the input is marked in gray and the SW of the system is marked with a dashed line. One can see that the condition in (1.16) is not fulfilled, i.e., the systems' SW does contain the inputs' SW, but their total area is about the same. But the condition expressed in (1.17) is fulfilled. In Fig. 3.12b, the input propagates in space, and in Fig. 3.12c we can see the effect of the grating. In Fig. 3.12d, the input virtually propagates backward. In Fig. 3.12f, we can see the input after passing through the systems aperture. As it appears on the output plane, one can see that all of the input frequencies have been transferred but need to be

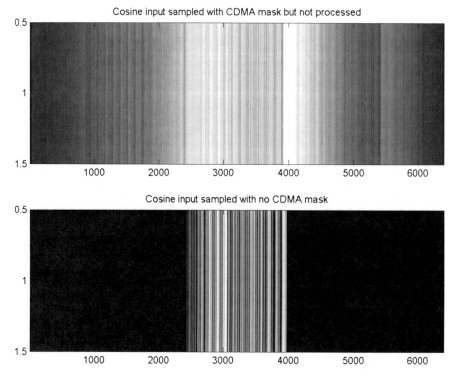

Fig. 3.9 (*Top*) One-dimensional sampled image from experiment with coding mask; (*bottom*) one-dimensional sampled image without coding mask. The image was taken from: J. Solomon, Z. Zalevsky and D. Mendlovic, "Geometrical Super Resolution by Code Division Multiplexing," Appl. Opt., 44, 32–40 (2005)

correctly decoded, in order to receive the original input with no distortion. The SW of the inputs was theoretically divided into three parts (base band frequencies, high negative frequencies, and high positive frequencies); these three parts were frequency modulated and transmitted through the system. This is very similar to the frequency division multiplexing approach (FDMA) in communications.

Retrieval of the image is done in a similar manner. The output of the system propagates in space (Fig. 3.12f). The signal is multiplied by the second grid (Fig. 3.12g), and if we virtually propagate, we obtain Fig. 3.12h. In the last figure, we receive the original input with its full resolution but with two ghost images. These ghost images limit the size of the original input image, as we must avoid overlapping of the original image with the ghost images.

3.2.2.2 Super Resolution Using CDMA

As has been shown above, traditional methods that have sacrificed image size for enhanced resolution used gratings. These actually implemented a setup in which the frequency is modulated by different parts of the image frequencies. It has been

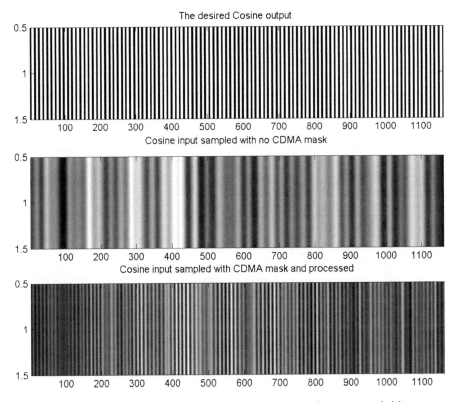

Fig. 3.10 (*Top*) Cosine grid, used as input; (*middle*) experimental output sampled by sensor; (*bottom*) experimental signal retrieved using CDMA superresolution. The image was taken from: J. Solomon, Z. Zalevsky and D. Mendlovic, "Geometrical Super Resolution by Code Division Multiplexing," Appl. Opt., 44, 32–40 (2005)

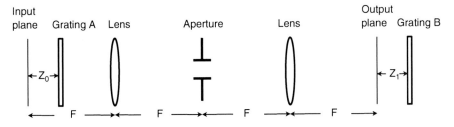

Fig. 3.11 Classical super resolution setup

shown that CDMA multiplexing almost achieves channel capacity and is superior to other methods such as FDMA [6].

This method is shown in Fig. 3.13. In Fig. 3.13a, one can see the original input in the Wigner space and the system's SW. The signal is Fourier transformed (or cosine transformed), Fig. 3.13b. The signal is multiplied by a coding mask, which expands its bandwidth. The signal is Fourier transformed again, Fig. 3.13d,

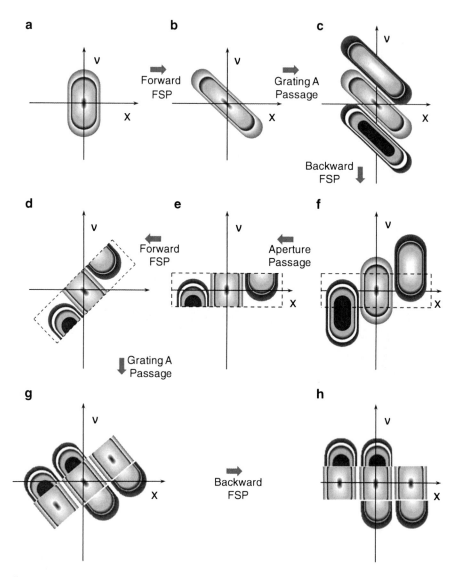

Fig. 3.12 Classical super resolution shown in the Wigner plane

and then multiplied by a grid, Fig. 3.13e. Notice that now different parts of the images produced due to the multiplication with the grid overlap. The CDMA grid must be constructed in a way that the different images will remain orthogonal. Finally, the signal passes through the system's aperture in Fig. 3.13f.

Retrieval of the image is done in a similar fashion. The image is multiplied by a grid, which is identical to the first one, and then Fourier transformed. Then it is multiplied by the CDMA grid and Fourier transformed again.

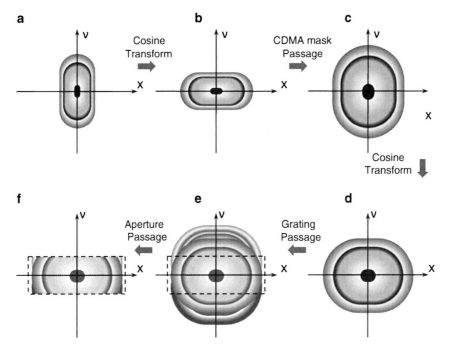

Fig. 3.13 CDMA super resolution shown in the Wigner plane

Optical setup for achieving this is shown in Fig. 3.14a. In order to implement the concept presented, one needs to multiply the optical image in the Fourier domain with the coding mask. Ordinary lenses perform this feat, but since they limit systems' resolution one needs to do this in a different manner. In Sect. 3.2.3, such a solution is presented.

Figure 3.14b illustrates the "computational path" used to calculate the output of the setup. The various optical and mathematical steps are:

- $u_0(x, 0) \rightarrow u_0(\mu, z_0^-)$ optical cosine transform
- $u_0(\mu, z_0^-) \rightarrow u_0(\mu, z_0^+)$ passing through coding mask
- $u_0(\mu, z_0^+) \rightarrow u_0(x, 2z_0^-)$ additional cosine transform
- $u_0(x, 2z_0^-) \rightarrow u_0(x, 2z_0^+)$ passing through Damman grating
- $u_0(x, 2z_0^+) \rightarrow u_0(\mu, (2F + 2z_0)^-)$ optical Fourier transform
- $u_0(\mu, (2F + 2z_0)^-) \rightarrow u_0(\mu, (2F + 2z_0)^-)\mathrm{rect}(\mu/\Delta\mu) = u_0(\mu, (2F + 2z_0)^+)$ passing through imaging systems' aperture
- $u_0(\mu, (2F + 2z_0)^+) \rightarrow u_0(x, 2F + 2z_0)$ optical Fourier transform

Image retrieval:

- $v_0(x_0^-) \rightarrow v_0(x_0^+)$ passing through Damman grating
- $v_0(x_0^+) \rightarrow v_0(\mu^-)$ Fourier transform
- $v_0(\mu^-) \rightarrow v_0(\mu^+)$ multiplication with coding mask
- $v_0(\mu^+) \rightarrow v_0(x)$ Fourier transform

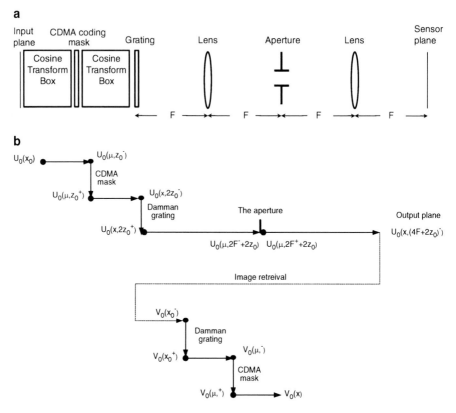

Fig. 3.14 (**a**) Full optical setup. (**b**) A flowchart illustrating the stages in the mathematical analysis of the optical setup and the computerized image retrieval. The image was taken from: J. Solomon, Z. Zalevsky and D. Mendlovic, "Super Resolution Using Code Division Multiplexing," Appl. Opt., 42, 1451–1462 (2003)

Note that the first grating should be moved only lightly for obtaining super resolved functionality of the proposed configuration.

3.2.2.3 Mathematical Analysis

Let us examine a simple situation, in which we want to enhance the resolution by a factor of 3 [7]. The coding mask composed of three subfunctions is given as follows:

$$G(v) = \sum_{n=-1}^{1} g_n(v - nv_0). \tag{3.25}$$

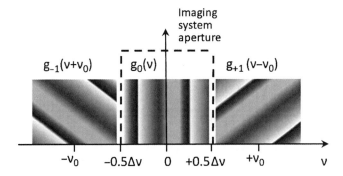

Fig. 3.15 Coding mask function broken into three subfunctions for each of the three ranges

This is shown graphically in Fig. 3.15.
g_i has the following properties:

$$\begin{cases} g_i(v) \cdot g_j(v) = 0, & i \neq j, \\ g_i(v) = 0, & i = -1, 0, 1 \end{cases} \forall v \notin \left[-\frac{v_0}{2}, \frac{v_0}{2} \right]. \tag{3.26}$$

In order to prove the above properties, a simple grating of three spectral orders will be chosen:

$$\text{grid}(v) = \delta(v) + \delta(v - v_0) + \delta(v + v_0). \tag{3.27}$$

The input image, marked by u_0, is convolved with the Fourier transform of the coding mask and multiplied by the grating before entering the imaging system. The system input image is marked as I.

$$I(x) = [u_0(x) * \text{CDMA}(x)] \cdot \text{grid}(x). \tag{3.28}$$

The systems input spectrum is derived from (3.28) producing:

$$
\begin{aligned}
I(v) &= \left[u_0(v) \sum_{n=-1}^{1} g_n(v - nv_0) \right] * [\delta(v) + \delta(v - 1) + \delta(v + 1)], \\
&= u_0(v) \sum_{n=-1}^{1} g_n(v - nv) + u_0(v - v_0) \sum_{n=-1}^{1} g_n(v - (n+1)v_0) \\
&\quad + u_0(v + v_0) \sum_{n=-1}^{1} g_n(v - (n-1)v_0).
\end{aligned}
\tag{3.29}
$$

Fig. 3.16 Cosine transform setup

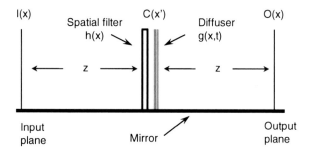

The inputs' spectrum is multiplied by the systems bandwidth (marked as $\text{rect}(v/\Delta v)$), and produces the systems' output, marked as O.

$$O(v) = I(v) \cdot \text{rect}\left(\frac{v}{\Delta v}\right)$$
$$= u_0(v)g_0(v) + u_0(v - v_0)g_{-1}(v) + u_0(v + v_0)g_1(v). \qquad (3.30)$$

Image retrieval is conducted as follows: The systems output is multiplied by an identical grating, producing R_1. This signal can be formulated in the frequency domain as follows:

$$R_1(v) = O(v) * \text{grid}(v) = u_0(v)g_0(v) + u_0(v - 2v_0)g_0(v - v_0) + u_0(v + 2v_0)g_0(v + v_0)$$
$$+ u_0(v - v_0)g_{-1}(v) + u_0(v - 2v_0)g_{-1}(v - v_0) + u_0(v)g_{-1}(v - v_0)$$
$$+ u_0(v + v_0)g_1(v) + u_0(v)g_1(v - v_0) + u_0(v + 2v_0)g_1(v + v_0).$$

$$(3.31)$$

Then this is multiplied with the coding mask, in the frequency domain. The result is marked as R_2:

$$R_2(v) = R_2(v) \cdot G(v) = u_0(v)g_0(v) + u_0(v)g_{-1}(v + v_0) + u_0(v)g_1(v - v_0),$$
$$= u_0(v) \cdot G(v) \xrightarrow[\text{downsampling}]{} u_0(v).$$

$$(3.32)$$

As one can see, after down sampling the filter, the output spectrum is identical to the original image spectrum, with no resolution decrease.

3.2.2.4 Optical Cosine Transform

Cosine transform can be implemented in incoherent illumination in a fashion which reminds the shearing interferometer [8]. A mirror, orthogonal to the input, is used to duplicate the input image. The optical setup (which used as a component in the CDMA super resolution scheme) is shown in Fig. 3.16.

Let us examine an input image, which is composed of a single dot, it is duplicated into two dots, by the mirror. These will interfere in a given distance z and produce an interference mask, which is similar to a cosine function. A single dot produced a cosine sequence. Due to the reversibility of ray optics, a cosine grid will produce a single dot in space. This hints that the setup actually cosine transforms the input. A more rigorous proof is shown below.

First, the output C is calculated in coherent illumination, while I is assumed as the input image:

$$
\begin{aligned}
C(x') &= \frac{\exp(jkz)}{j\lambda z} \exp\left(j\frac{k}{2z}x'^2\right) \int I'(x) \exp\left(j\frac{k}{2z}x^2\right) \exp\left(-j\frac{2\pi}{\lambda z}xx'\right) dx, \\
&= \frac{\exp(jkz)}{j\lambda z} \exp\left(j\frac{k}{2z}x'^2\right) \int \left[I(x)\exp\left(j\frac{k}{2z}x^2\right)\exp\left(-j\frac{2\pi}{\lambda z}xx'\right) \right. \\
&\quad \left. + I(x)\exp\left(j\frac{k}{2z}x^2\right)\exp\left(j\frac{2\pi}{\lambda z}xx'\right) \right] dx, \\
&= 2\frac{\exp(jkz)}{j\lambda z}\exp\left(j\frac{k}{2z}x'^2\right)\int I(x)\exp\left(j\frac{k}{2z}x^2\right)\mathrm{Cos}\left(\frac{2\pi}{\lambda z}xx'\right)dx. \quad (3.33)
\end{aligned}
$$

Now the output is expressed as a function of the space coordinate and time:

$$
C(x',t) = 2\frac{\exp(jkz)}{j\lambda z}\exp\left(j\frac{k}{2z}x'^2\right)\int I(x)\exp\left(j\frac{k}{2z}x^2\right)\mathrm{Cos}\left(\frac{2\pi}{\lambda z}xx'\right)dx. \quad (3.34)
$$

Let us assume incoherent illumination and calculate the intensity of the output:

$$
\begin{aligned}
|C(x')|^2 &= \frac{4}{\lambda^2 z^2}\iiint I(x_1,t)\bar{I}(x_2,t)\exp\left(j\frac{k}{2z}x_1^2\right)\exp\left(-j\frac{k}{2z}x_2^2\right)\mathrm{Cos}\left(\frac{2\pi}{\lambda z}x_1x'\right) \\
&\quad \mathrm{Cos}\left(\frac{2\pi}{\lambda z}x_2x'\right)dx_1dx_2dt, \\
&= \frac{4}{\lambda^2 z^2}\iint |I(x_1)|^2 \delta(x_1-x_2)\exp\left(j\frac{k}{2z}x_1^2\right)\exp\left(-j\frac{k}{2z}x_2^2\right)\mathrm{Cos}\left(\frac{2\pi}{\lambda z}x_1x'\right) \\
&\quad \mathrm{Cos}\left(\frac{2\pi}{\lambda z}x_2x'\right)dx_1dx_2 \\
&= \frac{4}{\lambda^2 z^2}\int |I(x)|^2\mathrm{Cos}^2\left(\frac{2\pi}{\lambda z}xx'\right)dx \\
&= \frac{2}{\lambda^2 z^2}\int |I(x)|^2 dx + \frac{2}{\lambda^2 z^2}\int |I(x)|^2\mathrm{Cos}\left(\frac{4\pi}{\lambda z}xx'\right)dx, \\
&= A_0 + \frac{2}{\lambda^2 z^2}\int |I(x)|^2\mathrm{Cos}\left(\frac{4\pi}{\lambda z}xx'\right)dx.
\end{aligned}
$$

$$(3.35)$$

This proved that this setup can be used in order to cosine transform the input signal (which is identical to the Fourier transform for real input images). If a diffuser is used, this setup can be used to cosine transform the results again, this is mathematically proven below. Let C denote the output of the first cosine transform, and F will be the effective result of the mirror used on C after the coding mask, h, and the diffuser, g.

$$F(x',t) = C(x',t)h(x',t)g(x',t) + C(-x',t)h(-x',t)g(-x',t). \tag{3.36}$$

The input after passing through the coding mask and the diffuser will be the free space propagation for a distance of z. This is expressed as O.

$$O(x_t',t) = \frac{\exp(jkz)}{j\lambda z}\exp\left(j\frac{k}{2z}x'^2\right)\int F(x,t)\exp\left(j\frac{k}{2z}x^2\right)\exp\left(-j\frac{2\pi}{\lambda z}xx'\right)dx. \tag{3.37}$$

The intensity measured by the sensor will be:

$$|O(\bar{x})|^2 = \frac{4}{\lambda^2 z^2}\int\int\int F(x',t)\bar{F}(x'',t)\exp\left(j\frac{k}{2z}x'^2\right)\exp\left(-j\frac{2\pi}{\lambda z}xx'\right)$$
$$\times \exp\left(-j\frac{k}{2z}x''^2\right)\exp\left(j\frac{2\pi}{\lambda z}xx''\right)dx'dx''dt. \tag{3.38}$$

Let us integrate the only terms affected over time:

$$\int F(x',t)\overline{F(x'',t)}dt = h(x')h(x'')\langle g(x',t)g(x'',t)\rangle\langle c(x',t)c(x'',t)\rangle$$
$$+ h(x')h(-x'')\langle g(x',t)g(-x'',t)\rangle\langle c(x',t)c(-x'',t)\rangle$$
$$+ h(-x')h(x'')\langle g(-x',t)g(x'',t)\rangle\langle c(-x',t)c(x'',t)\rangle$$
$$+ h(-x')h(-x'')\langle g(-x',t)g(-x'',t)\rangle\langle c(-x',t)c(-x'',t)\rangle. \tag{3.39}$$

Due to the physical nature of the diffuser, this can be expressed as:

$$\int F(x',t)\overline{F(x'',t)}dt = h(x')h(x'')\delta(x'-x'')\langle c(x',t)c(x'',t)\rangle$$
$$+ h(x')h(-x'')\delta(x'+x'')\langle c(x',t)c(-x'',t)\rangle$$
$$+ h(-x')h(x'')\delta(x'+x'')\langle c(-x',t)c(x'',t)\rangle$$
$$+ h(-x')h(-x'')\delta(x'-x'')\langle c(-x',t)c(-x'',t)\rangle. \tag{3.40}$$

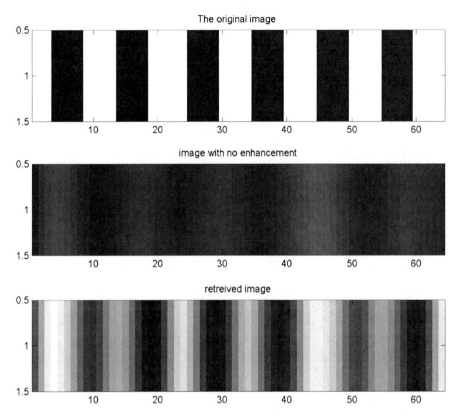

Fig. 3.17 Simulated input and output: (*top*) the original input image used; (*middle*) the output of the system with no resolution enhancement; (*bottom*) the output of the CDMA super resolving. The image was taken from: J. Solomon, Z. Zalevsky and D. Mendlovic, "Super Resolution Using Code Division Multiplexing," Appl. Opt., 42, 1451–1462 (2003)

Equation 3.28 can now be formulated according to (3.30) as:

$$
\begin{aligned}
|O(\bar{x})|^2 &= \frac{4}{\lambda^2}z^2 \int \Big[\big\{ |h(x)|^2|c(x)|^2 + |h(-x)|^2|c(-x)|^2 \big\} \\
&\quad + \big\{ |h(-x)|^2|c(x)|^2 + |h(x)|^2|c(-x)|^2 \big\} \exp^2\Big(-j\frac{2\pi}{\lambda z}xx'\Big) \Big] \mathrm{d}x', \\
&= B_0 + \frac{4}{\lambda^2 z^2} \int \Big[\big\{ |h(x)|^2|c(x)|^2 + |h(-x)|^2|c(-x)|^2 \big\} \exp\Big(-j\frac{4\pi}{\lambda z}xx'\Big) \Big] \mathrm{d}x', \\
&= B_0 + \frac{4}{\lambda^2 z^2} \int \Big[\big\{ |h(x)|^2|c(x)|^2 \big\} \mathrm{Cos}\Big(\frac{4\pi}{\lambda z}xx'\Big) \Big] \mathrm{d}x'. \quad (3.41)
\end{aligned}
$$

This is the intensity measured by the sensor. This proved that this setup can be used in order to cosine transform the input signal (which is identical to the Fourier

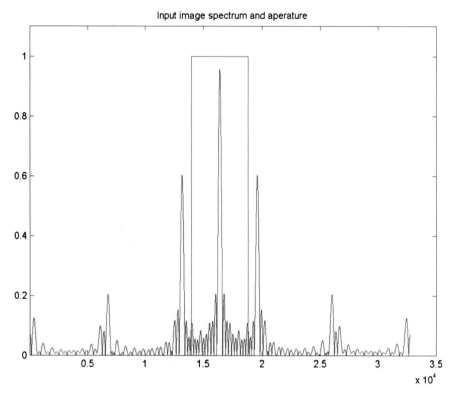

Fig. 3.18 Input image spectrum and systems' bandwidth (marked by *rectangle*). The image was taken from: J. Solomon, Z. Zalevsky and D. Mendlovic, "Super Resolution Using Code Division Multiplexing," Appl. Opt., 42, 1451–1462 (2003)

transform for real input images), multiply it by a mask and then cosine transform the input again, i.e., effectively convolving the signal with the Fourier transform of the mask.

3.2.3 Computer Simulations

In order to verify the method presented in Sect. 3.2.2 for CDMA super resolution, a MATLAB simulation was constructed. An input image of 64 by 64 pixels was chosen. A coding mask was constructed in a special manner. It was done in a way that after convolving with a grating, the different orders of the coding masks would not overlap, i.e., the transparent parts will not overlap.

In the simulation, the input was cosine transformed, multiplied with the coding mask and cosine transformed again. It was multiplied by a grating and then it was transferred through a system with limited bandwidth. This simulated setup is described in Sect. 3.2.2 according to the computational steeps presented in Fig. 3.14b.

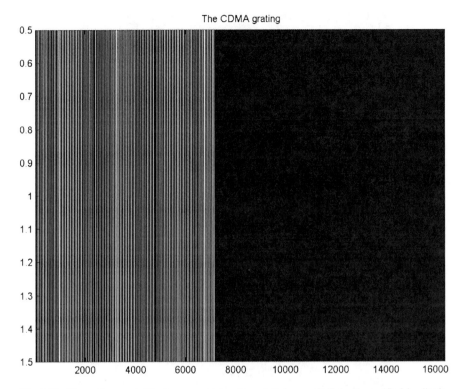

Fig. 3.19 The coding mask. The image was taken from: J. Solomon, Z. Zalevsky and D. Mendlovic, "Super Resolution Using Code Division Multiplexing," Appl. Opt., 42, 1451–1462 (2003)

The input used was a simple binary grating shown in Fig. 3.17 (top). Systems bandwidth was chosen so as not to enable the transmission of the input image resolution. The spectrum of the input image and the systems bandwidth are shown in Fig. 3.18. The coding mask chosen is shown in Fig. 3.19.

The output of the system is presented in Fig. 3.17: In the middle image, the output is shown for the system without applying the super resolution method. One can see a complete loss of image resolution. The bottom image shows the input after construction using CDMA super resolution; one can see that the resolution was almost completely reconstructed.

3.2.4 Experimental Results

In order to verify the method presented in Sect. 3.2.2 for performing optical cosine transform, an experimental setup was constructed. The setup was only half of the setup in Fig. 3.16, i.e., it was composed of an input image, a mirror, and a sensor for recording the results. The input used was a grid of 100 lines per millimeter.

Fig. 3.20 Output of cosine setup with coherent illumination. The image was taken from: J. Solomon, Z. Zalevsky and D. Mendlovic, "Super Resolution Using Code Division Multiplexing," Appl. Opt., 42, 1451–1462 (2003)

Fig. 3.21 Output of the cosine setup for incoherent illumination but without a mirror. The image was taken from: J. Solomon, Z. Zalevsky and D. Mendlovic, "Super Resolution Using Code Division Multiplexing," Appl. Opt., 42, 1451–1462 (2003)

First, the output of the setup was recorded for coherent illumination, and is presented in Fig. 3.20. Then the setup was used with incoherent illumination but without a mirror, the output was just a blurred spot, as shown in Fig. 3.21. Finally, the setup was tested with incoherent illumination and a mirror, among the blurred spot, fine fringes appeared as outlined in Fig. 3.22. Notice that the line spacing is

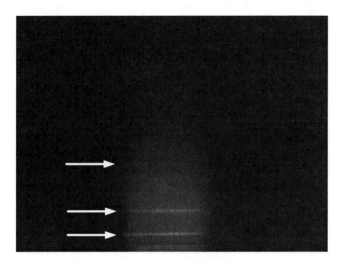

Fig. 3.22 Output of the cosine setup with incoherent illumination, *arrows* emphasize fringe location. The image was taken from: J. Solomon, Z. Zalevsky and D. Mendlovic, "Super Resolution using Code Division Multiplexing," Appl. Opt., 42, 1451–1462 (2003)

identical to the result received with the coherent illumination. This is obvious since the far field approximation of the input and the cosine transform will produce the same spatial frequency of the input grating. Notice that the lines do not coincide since the coherent beam that illuminated the input pattern for the coherent case had a little elevation angle. Thus, in the far field, this angle caused a shift in the received fringes. Obviously, this angle had no effect for the incoherent illumination case.

References

1. Solomon, J., Zalevsky, Z., Mendlovic, D.: Geometrical super resolution using code division multiplexing. Appl. Opt. **42**, 32–40 (2005)
2. Viterbi, A.J.: CDMA Principles of Spread Spectrum Communication. Addition-Wesley, Reading, NA (1995)
3. Goodman, J.: Introduction to Fourier Optics, International 2nd edn, pp. 101–104. McGraw-Hill, Singapore (1996)
4. Taub, H., Schilling, D.L.: Principles of Communication Systems, 2nd edn. McGraw-Hill, New York (1986)
5. Lukosz, W.: Optical systems with resolving powers exceeding the classical limit. J. Opt. Soc. Am. **56**, 1463–1472 (1966)
6. Viterbi, A.J.: CDMA, Principles of Spread Spectrum Communication, Addison-Wesley, New York (1995)
7. Solomon, J., Zalevsky, Z., Mendlovic, D.: Super resolution using code division multiplexing. Appl. Opt. **42**, 1451–1462 (2003)
8. Mendlovic, D., Zelavsky, Z., Konforti, N.: Joint transform correlator with incoherent output. J. Opt. Soc. Am. **A11**, 3201–3205 (1994)

Chapter 4
Techniques Utilizing Diffractive Masks Having Structures with a Period Non-Limited Randomness

Alex Zlotnik, Zeev Zalevsky, Amikam Borkowski, David Sylman, Vicente Micó, Javier García, and Bahram Javidi

4.1 Geometrical Super Resolved Imaging Using Nonperiodic Spatial Masking

4.1.1 Introduction

In this section, an approach to overcome the geometrical resolution limitations is presented [1]. In a sensor array, the resolution is determined by the spatial density of the pixels on the image plane, i.e., the number of pixels per unit area, and by the geometrical point spread function (PSF) which characterizes the spatial responsivity of each pixel. The limitation which is related to the spatial density of the pixels can easily be removed by a procedure called microscanning and interlacing, i.e., adding many low-resolution images, each taken at a slightly different small geometrical shifts [2].

The under sampling aspect of the geometrical limitation which is solved by applying microscanning and by interlacing several low-resolution images into one unified image with larger number of pixels is not the main problem that this section aims to address. After the interlacing, one can obtain an image having a lot of pixels but which is still blurred due to the large pixels that are used to create this image [3, 4]. De-blurring this image and improving its resolution to correspond to the pitch of the sampling grid of the interlaced image is the main issue addressed. In other words, it is meant to overcome the PSF created by the spatial responsivity of each pixel after the generation of the interlaced unified image.

The novelty of the described approach is in adding a priori knowledge which allows proper matrix inversion and true extraction of super resolved information rather than extrapolation as applied in many other digital super resolved approaches.

Z. Zalevsky (✉)
School of Engineering, Bar-Ilan University, Ramat-Gan, Israel
e-mail: zalevsz@macs.biu.ac.il

Z. Zalevsky (ed.), *Super-Resolved Imaging: Geometrical and Diffraction Approaches*,
SpringerBriefs in Physics, DOI 10.1007/978-1-4614-0833-8_4,
© Springer Science+Business Media, LLC 2011

The method is applied over an interlaced image that is generated after microscanning procedure. The microscanning is a procedure such that the mask and the object should be moving together in respect to the detector (this can be obtained if the scanning mirror mechanism is positioned between the mask and the detector rather than between the object and the mask). Extra optical hardware includes a binary transparency mask located in the intermediate imaging plane of the optics. The mask adds the required a priori knowledge for the matrix inversion. The described technique aims to obtain geometrical super resolution, i.e., it assumes that the diffraction limitation related to the F-number of the optics can be ignored. Therefore, in diffraction limited system having F-number of 1, the maximal geometrical resolution that the proposed approach should aim for is the optical wavelength (i.e., about half a micron).

Additional related works that use some form of the movement of the camera or modulation of the pupil function are described in refs. [5–8].

4.1.2   Theoretical Analysis

4.1.2.1   General Description

The proposed idea includes positioning one of the three possible masks in the intermediate image plane of the imaging optics. The mask contains either a frame of zero intensities at the borders of the field of view or a binary mask of large random pixels having the size of the original resolution of the imager or a binary mask of small random pixels having the size of the intended high resolution. Each one of those three masks actually imposes conditions of zero energy (or constant energy) over the high-resolution image before it is being sampled by the detector (and therefore blurred due to the shape of the sampling pixel). Knowing those conditions allows extracting the high-resolution information despite the significant spatial degradation by the detector (i.e., having large pixels with low resolution).

A convenient way to observe the problem of resolution enhancement is by looking at each pixel in the sampling array as a subpixel matrix functioning as an averaging operator. Therefore, for each shift of the camera, some new subpixels are added and some are removed from the averaging. This is equivalent to the convolution of a super resolved image and a function having the spatial shape of a pixel. This convolution can be farther described as a set of linear equations $A\hat{x} \approx b$, where $\hat{x}$ represents the super resolution image translated to a vector, b represents the interlaced discrete image constructed as spread vector, and A represents a convolution action caused due to the spatially extended shape of the pixels. The naive answer to this problem is to inverse the convolution matrix and to extract the estimator $\hat{x}$ as $\hat{x} = bA^{-1}$. Since there is no existence of a solution, one may use the method of reducing the least square error $\|Ax - b\|^2$, such that $\hat{x} = (A^{T}A)^{-1}A^{T}b$ (i.e., a pseudoinverse) or using Tikhonov's regularization and obtaining $\hat{x} = (A^{T}A + \alpha^2 I)^{-1}A^{T}b$ [9].

Note that the resolution improvement algorithm is applied after performing microscanning which increases the effective number of low-resolution pixels in the captured image. The suggested approach allows improvement of resolution by an order of magnitude in every axis as to be demonstrated, i.e., a low-resolution image that is blurred due to the large pixels of the detector is being microscanned. The microscanning factor equals to the super resolution factor to be obtained. The outcome is a blurred image with number of pixels corresponding to the desired resolution. The blurring is obtained due to the originally large geometry of the pixels in the detector. If one knows some a priori initial conditions which can be at the border of the image or at random locations along the image, then the de- convolution operation is feasible. Basically, the convolution operation (the blurring) may be expressed as a linear operation applied over the image, i.e., $Ax = b$ where A is the matrix of blurring, x are the unknown variables that are the pixels of the high-resolution image which are to be extracted, and b is the captured low-resolution image. In an image of size $(M - R - 1) \times (N - R - 1)$ where R stands for the size of the blurring kernel one usually has $N \times M$ unknown variables which make the problem of inversion impossible unless it is based upon extrapolation. Due to the mask positioned in the intermediate plane, the number of variables is reduced (by blocking the light at certain spatial positions). Therefore, there will be $(M - (R - 1)) \times (N - (R - 1))$ equations as well as unknown variables. By posing zeros at the edges of the field of view or at random positions in the image, the number of variables is reduced. This allows simple extraction of the real super resolved information without "speculations" (i.e., without extrapolation).

Note that since R is small (say 8 in comparison to N or M which are 1,000 or more), the "loss" of spatial information within the field of view or the reduction of the field of view (when the zeros are at the borders) is negligible.

Therefore, the purpose of the spatial mask is as follows: addition of known information in order to improve the ratio between the number of equations and the number of unknown variables. The mask blocks some of the original information but adds new known information allowing the precise inversion of the matrix A. As previously mentioned, there are three possibilities:

- Mask blocking the edges of the field of view. The advantage of this mask is that it is simple for fabrication and actually does not have to be positioned in the intermediate plane. Instead, the optics itself can be modified in order to realize this. Note also that this mask does not contain zeros. It may just have uniform gray levels. The disadvantage is that the inversion is good at the proximity of the border and it has larger accumulated error noise toward the center of the image.
- Random binary mask with large pixels. The a priori knowledge is randomly spread along the entire image, therefore, the error is not accumulated. Since the pixels are large (have the size of the low-resolution image), more energy and more spatial information is lost. One solution is to use two masks each providing blocking in different spatial positions. Then, by time multiplexing (which is applied in addition to the microscan), one may capture two images each with different mask and recover the high resolution over the full field of view by combining the two super resolved results. The optical implementation of such a time multiplexing can be obtained by positioning spatial light modulator (SLM) in the intermediate image plane of the optical system and displaying two different random sequential binary distributions.

- Random binary mask with small pixels. The a priori knowledge is also randomly spread along the entire image. Therefore, the error is not accumulated. Since the pixels are small (the size of the super resolved image), less energy and less spatial information is lost. The final result will be the super resolved image multiplied by the random mask. Another mask is needed to recover the missing information. In this case, the spatial information is lost but rather there are some areas in the field of view where there is no super resolution. Therefore, the use of the second mask (temporal multiplexing) is not crucial (but may be used) and this is the main advantage of this approach.

4.1.2.2 Algorithm Description

General Outline

As was mentioned previously, the interlaced image can be modeled by a convolution operation:

$$y[m, n] = g[m, n] * u[m, n] = \sum_{k=0}^{R-1} \sum_{l-0}^{R-1} g[k + m, l + n] u[k, l], \qquad (4.1)$$

where the functions u and y represent the original and the blurred images, respectively. The smoothed kernel g is referred to the spatial shape of the pixel with a size of R.

The original and blurred images, i.e., the functions u and y, are assumed to be defined by a compact set of linear equations:

$$\underline{b} = \underline{\underline{A}} \underline{x}, \qquad (4.2)$$

where the vector $\underline{x}$ contains nonnegative pixel values, ordered column-wise, of the discrete image that corresponds to u. The vector $\underline{b}$ contains the pixel values, ordered column-wise, of the blurred and sampled images associated with $\underline{x}$. The matrix $\underline{\underline{A}}$ is the blurring matrix obtained by sampling the summing operator. Every pixel in the blurred image y represents a linear multiplication between $\underline{b}$ and $\underline{x}$. R-doing this operation for each pixel, one can obtain many different row vectors. Combining all of these vectors creates the matrix $\underline{\underline{A}}$. In the 1-D case, this matrix has a Toeplitz Block form. In the 2-D case, the process is much more complicated, and it requires the integration of many Toeplitz blocks in a Toeplitz block structure using the Kronecker product. Matrices with this structure are referred to as block Toeplitz form with Toeplitz blocks (BTTB) matrices.

Let us now demonstrate the operation principle of the algorithm with a concrete example. Let:

$$u = \begin{bmatrix} u_{00} & \cdots & u_{04} \\ \vdots & & \vdots \\ u_{20} & \cdots & u_{24} \end{bmatrix}$$

be a 3×5 pixels image, such that x will be ordered column-wise as:

$$x = \begin{bmatrix} u_{00} \\ \vdots \\ u_{04} \\ u_{10} \\ \vdots \\ u_{14} \\ u_{20} \\ \vdots \\ u_{24} \end{bmatrix}.$$

The smoothing kernel $g = \frac{1}{4}\begin{bmatrix} 1 & 1 \\ 1 & 1 \end{bmatrix}$ having the size of 2×2 will perform four longitudinal and two lateral steps. First, let us start by describing the longitudinal steps of the first row of the kernel. The first step in this row is to multiply the u_{00}, u_{01} terms and sum them up. The row-wise vector will be shown as $\begin{bmatrix} 1 & 1 & 0 & 0 & 0 \end{bmatrix}$. The next step in this row is to multiply the u_{01}, u_{02} terms and to sum them up. The row-wise vector will be shown as $\begin{bmatrix} 0 & 1 & 1 & 0 & 0 \end{bmatrix}$ and so on. After gathering all these vectors to a matrix, a Toeplitz block will be obtained (denoted by A_2):

$$A_2 = \frac{1}{4}\begin{bmatrix} 1 & 1 & 0 & 0 & 0 \\ 0 & 1 & 1 & 0 & 0 \\ 0 & 0 & 1 & 1 & 0 \\ 0 & 0 & 0 & 1 & 1 \end{bmatrix}.$$

The other kernel's row and the lateral steps can be described as a Toeplitz block as well where every term in this matrix is represented as the first Toeplitz block A_2 that have been obtained. Since this kernel includes two rows representing two longitudinal movements and two lateral movements, one can obtain a new Toeplitz block $A_1 = \begin{bmatrix} 1 & 1 & 0 \\ 0 & 1 & 1 \end{bmatrix}$. By replacing every term in this block with the first Toeplitz block A_2 (using the Kronecker product), the final matrix takes next form:

$$\underline{\underline{A}} = A_1 \otimes A_2^{\mathrm{T}} = \frac{1}{4}\left[\begin{array}{ccccc|ccccc|c} 1 & 1 & 0 & 0 & 0 & 1 & 1 & 0 & 0 & 0 & \\ 0 & 1 & 1 & 0 & 0 & 0 & 1 & 1 & 0 & 0 & 0 \\ 0 & 0 & 1 & 1 & 0 & 0 & 0 & 1 & 1 & 0 & \\ 0 & 0 & 0 & 1 & 1 & 0 & 0 & 0 & 1 & 1 & \\ \hline & & & & & 1 & 1 & 0 & 0 & 0 & 1 & 1 & 0 & 0 & 0 \\ & & 0 & & & 0 & 1 & 1 & 0 & 0 & 0 & 1 & 1 & 0 & 0 \\ & & & & & 0 & 0 & 1 & 1 & 0 & 0 & 0 & 1 & 1 & 0 \\ & & & & & 0 & 0 & 0 & 1 & 1 & 0 & 0 & 0 & 1 & 1 \end{array}\right].$$

Every row in this matrix has 15 variables, as expected, and it represents exactly the linear multiplication with the $\underline{x}$ column vector of the image, therefore the $\underline{b}$ column vector for the blurred image:

$$\underline{b} = \begin{bmatrix} y_{00} \\ \vdots \\ y_{03} \\ y_{10} \\ \vdots \\ y_{13} \end{bmatrix}$$

has the same length as the number of rows in the matrix $\underline{\underline{A}}$.

By multiplying the vector b by the inverse matrix A^{-1}, it is possible to recover the original high-resolution vector x. Since the matrix has more rows than columns (i.e., there are more variables than equations), the matrix is not square and therefore it is not invertible. One possibility is to use pseudoinverse, i.e., to use the minimum mean square error of $\|Ax - b\|^2$:

$$\|Ax - b\|^2 = (Ax - b)^{\mathrm{T}}(Ax - b) = (Ax)^{\mathrm{T}}(Ax) - b^{\mathrm{T}}Ax - (Ax)^{\mathrm{T}}b + b^{\mathrm{T}}b. \quad (4.3)$$

Setting the gradient to zero yields:

$$\frac{\mathrm{d}}{\mathrm{d}x}\left[\|Ax - b\|^2\right] = 2A^{\mathrm{T}}A\hat{x} - 2A^{\mathrm{T}}b = 0. \quad (4.4)$$

Therefore:

$$\hat{x}(A^{\mathrm{T}}A) = A^{\mathrm{T}}b \Rightarrow \hat{x} = (A^{\mathrm{T}}A)^{-1}A^{\mathrm{T}}b = A_{\mathrm{p}}^{-1}b, \quad (4.5)$$

where A_{p}^{-1} is the pseudoinverse and $\hat{x}$ is the estimation for the unknown variables x. However, since the original image contains noise (quantization noise, shot noises, thermal noises, etc.), the solution of $\hat{x} = A_{\mathrm{p}}^{-1}b$ will not produce good results and may even be unstable mathematically. Therefore, additional regularization term is added to avoid this problem and to reduce sensitivity to existing noise. Tikhonov regularization term $\|\Gamma x\|^2$ is used to find a smooth solution:

$$\begin{aligned} \|Ax - b\|^2 + \|\Gamma x\|^2 &= (Ax - b)^{\mathrm{T}}(Ax - b) + (\Gamma x)^{\mathrm{T}}(\Gamma x), \\ &= (Ax)^{\mathrm{T}}(Ax) - b^{\mathrm{T}}Ax - (Ax)^{\mathrm{T}}b + b^{\mathrm{T}}b + (\Gamma x)^{\mathrm{T}}(\Gamma x). \end{aligned} \quad (4.6)$$

Setting the gradient to zero leads to:

$$\frac{\mathrm{d}}{\mathrm{d}x}\left[\|Ax - b\|^2 + \|\Gamma x\|^2\right] = 2A^{\mathrm{T}}A\hat{x} - 2A^{\mathrm{T}}b + 2\Gamma^{\mathrm{T}}\Gamma\hat{x} = 0, \quad (4.7)$$

$$\hat{x}\left(A^{\mathrm{T}}A + \Gamma^{\mathrm{T}}\Gamma\right) = A^{\mathrm{T}}b \ \Rightarrow\ \hat{x} = \left(A^{\mathrm{T}}A + \Gamma^{\mathrm{T}}\Gamma\right)^{-1}A^{\mathrm{T}}b. \tag{4.8}$$

If one assumes that $\Gamma = \alpha I$, where I is the unity matrix, then the obtained solution is:

$$\hat{x} = \left(A^{\mathrm{T}}A + \alpha^2 I\right)^{-1}A^{\mathrm{T}}b, \tag{4.9}$$

while α can be estimated using various approaches such as the Bayesian interpretation, the discrepancy principle, the cross validation, the L-curve method, the unbiased predictive risk estimator, or the leave-one-out cross-validation approach [10].

By applying the mask, the information added by the mask improves the optimization process. Each transparent pixel of the mask does not affect the calculation procedure, and it is averaged as before due to the large pixel's size in the detector which is the low geometrical resolution. Each opaque pixel in the mask eliminates the information at that point, and it is not being summed with the rest of the pixels. In fact, each time a "zero" pixel appears, an equation line is added to the matrix $\underline{A}$, composed of zeroes and single "1" values, and add the value "0" in the outcome vector $\underline{b}$:

$$\underline{A} = \frac{1}{4}\begin{bmatrix} 1 & 0 & \cdots & & & & & & & & & \cdots & 0 \\ 0 & 0 & 1 & 0 & \cdots & & & & & & & \cdots & 0 \\ 0 & & \cdots & & 0 & 1 & 0 & & & \cdots & & & 0 \\ 0 & & & & 0 & 1 & 0 & & & \cdots & & & 0 \\ 0 & & \cdots & & & & 0 & 1 & 0 & & \cdots & & 0 \\ 0 & & & & & & \cdots & 0 & 1 & 0 & 0 & & \\ 0 & & & & & & \cdots & & 0 & 1 & 0 & & \\ & \vdots & & \vdots & & \vdots & & \vdots & & \vdots & & & 0 \\ 1 & 1 & 0 & 0 & 0 & 1 & 1 & 0 & 0 & 0 & 0 & 0 & 0 & 0 & 0 \\ 0 & 1 & 1 & \ddots & & 0 & 1 & 1 & & & & & \\ & & 1 & 1 & 0 & & \ddots & 1 & 1 & & & \\ & & & 1 & 1 & 0 & 0 & 0 & 1 & 1 & & & \\ \vdots & & & & 1 & 1 & 0 & 0 & 0 & 1 & 1 & & \\ & & & & 1 & 1 & \ddots & & 0 & 1 & 1 & \\ & & & & & 1 & 1 & 0 & & \ddots & 1 & 1 & 0 \\ 0 & 0 & 0 & 0 & 0 & 0 & 0 & 0 & 1 & 1 & 0 & 0 & 0 & 1 & 1 \end{bmatrix}, \quad \underline{b} = \begin{bmatrix} 0 \\ \vdots \\ 0 \\ y_{00} \\ \vdots \\ y_{03} \\ y_{10} \\ \vdots \\ y_{13} \end{bmatrix}.$$

In fact, the matrix $\underline{A}$ may become a square matrix and even one that contains more rows than columns, making the system more constrained. Thus, it is possible to implement the inverse algorithm with the combination of the Tikhonov regularization, getting substantially more accurate result.

In summary, the proposed approach includes deploying a hardware mask which adds important a priori information and capturing a set of microscanned images. The interlaced image generates a blurred image but with the number of pixels following the required super resolving factor. Then, by applying Tikhonov regularization and inverting the extended (i.e., modified) matrix A, one may extract the super resolved information.

Computational Requirements

The size of the matrix is $(M \times N) \times [(M - (R - 1)) \times (N - (R - 1))]$. The matrix with the "known elements," which is only a diagonal matrix with the size of $(M \times N) \times (M \times N)$ should be added to matrix A, creating a very large matrix, whose size is approximately $(M \times N) \times 2(M \times N)$. The use of Tikhonov regularization requires triple matrix multiplication for the matrix A and calculation of the invertible matrix $(A^{\mathrm{T}} A)^{-1}$, which is a time-consuming manipulation.

In order to increase the processing speed, the computations are not done over the entire image at once but rather each time the computations are performed over a segment of the image. The result is used for the computation of the next segment.

One can dramatically reduce the computation requirements if some a priori information on the object shape or location is known. For example, if the task is to perform sort of digital zooming in which one wishes to improve the geometrical resolution only in a limited and specified spatial region, then the region of interest (ROI) is only a well-known number of neighbored pixels. This can reduce the number of calculations involved in the Fourier/inverse Fourier transform computations. That is, in some problems, the selection of proper ROI decreases the computational complexity.

4.1.3 Experimental Investigation

Schematic sketch of the experimental setup is presented in Fig. 4.1 (the experimental setup is part of a *DarkField* wafer inspection system). The experimental system contains CMOS camera of Phantom v10 with $2,400 \times 1,800$ pixels capable of producing images at a frame rate of 480 fps, with focal lengths of BFL = EFL = 30 mm and a vibration stage with the precision of 0.5 µm. The active area used for the image capturing is $1,280 \times 800$ pixels. The camera has pixels with a size of 11.5×11.5 µm. In the experiments, each group of eight pixels is binned (i.e., generated pixels of 92×92 µm). Super resolved image is therefore of original resolution corresponding to pixels with a size of 11.5×11.5 µm. The camera had eight bits of dynamic range and thus the captured images had gray levels varying from 0 to 255. The optical system has different horizontal and vertical magnification factors. In the horizontal axis, the magnification is 2.3 (generating foot print of 5 µm over the wafer plane). In the vertical axis, the magnification is 5.75 (generating foot

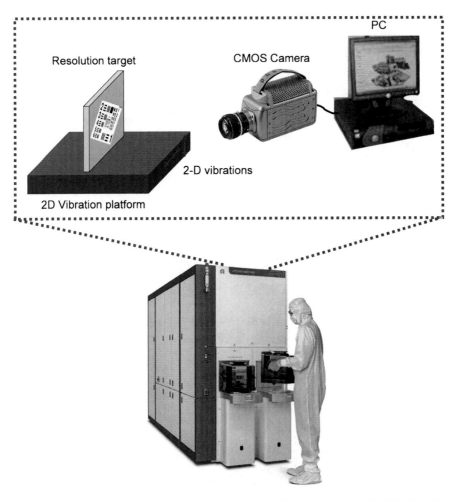

Fig. 4.1 Schematic sketch of the experimental setup which is part of the DarkField wafer inspection system of Applied Materials Ltd. The image was taken from: A. Borkowski, Z. Zalevsky and B. Javidi, "Geometrical Super Resolved Imaging Using Non periodic Spatial Masking," JOSA A 26, 589–601 (2009)

print of 2 μm over the wafer plane). In the experiments, the stage is aimed to generate relative movement of the resolution target/object and the random mask in comparison to the detector. The stage allowed moving the object in steps of 11.5 μm, then stopping and allowing the system to capture an image. As mentioned, the effective size of the pixels in the system is 92 μm and this is the geometrical limit for the resolution to be achieved. Thus, the movement of 11.5 μm is a subpixel movement equivalent to steps of 1/8 of the size of the original pixel (92 μm).

Figure 4.2a shows a high-resolution reference image captured by a camera. The low-resolution image is seen in Fig. 4.2b. The reduction of resolution is by a factor of 8 in every axis. Figure 4.2c shows the experimentally obtained image.

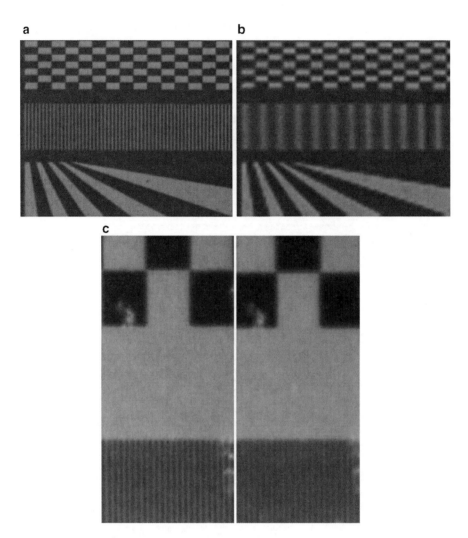

Fig. 4.2 (a) High-resolution reference image. (b) Low-resolution image (without super resolution). (c) Experimentally obtained image where in the left side the image is obtained after performing regular microscanning procedure and in the right side it is without it. The image was taken from: A. Borkowski, Z. Zalevsky and B. Javidi, "Geometrical Super Resolved Imaging Using Non periodic Spatial Masking," JOSA A 26, 589–601 (2009)

The image in the left side was obtained after performing regular microscanning procedure and in the right side it is without it. The improvement is visible but it is less than a factor of 2. Figure 4.2c was generated by shifting the object in steps of 11.5 μm and interlacing the 64 captured images into a unified image with higher sampling resolution.

In the following figures, the images were captured using previously described experimental platform. An example of the random blocking mask having 50% of

Fig. 4.3 An example of the random mask with 50% blocking. The image was taken from: A. Borkowski, Z. Zalevsky and B. Javidi, "Geometrical Super Resolved Imaging Using Non periodic Spatial Masking," JOSA A 26, 589–601 (2009)

blocking is presented in Fig. 4.3. Figure 4.4a–c presents three super resolved reconstructions with three types of masks described in Sect. 4.1.2.1 – (a) field of view limiter, (b) random low-resolution mask, and (c). random high-resolution mask.

One may see that the reconstruction result is similar to the original high-resolution image. In all three approaches, noise is added and the outcome is proven to have reduced noise sensitivity. In this simulation, the reconstructed images were obtained while applying the time multiplexing such that no black regions are remained in the reconstruction. In the case of Fig. 4.4a, the field of view blocking mask blocked seven high-resolution pixels (11.5 μm) from each side of the field of view.

Figure 4.5 shows the masked image (left) and the blurred and masked images (right) as they appear in the two cases where the random mask is applied. Figure 4.5a and b is the cases of high-resolution and low-resolution random masks correspondingly. The images were obtained through reconstruction of a single image, without applying the time multiplexing approach which allows reconstruction over the full field of view.

Next the sensitivity to noise is examined. Figure 4.6a and b shows the reference high-resolution image without noise and with embedded noise having the standard deviation of 20 gray levels, respectively. The low resolution is shown in Fig. 4.6c. The reconstruction with first kind mask, i.e., by blocking the edges of the field of view, is seen in Fig. 4.6d. Figure 4.6e and f shows reconstructed image with a second and third kind of random masks, respectively. One may see that despite the noise most of the information is recovered in all three approaches due to the applied process of regularization. Next, the sensitivity of the quality of reconstruction to

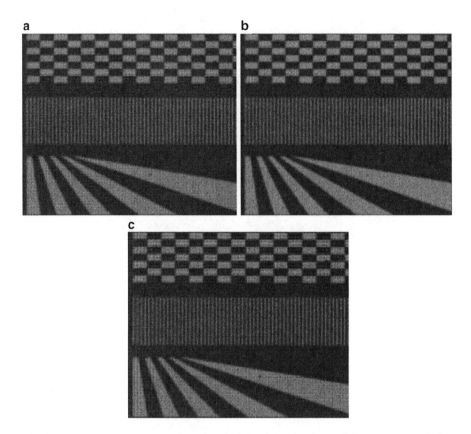

Fig. 4.4 Super resolved reconstruction using (**a**) field of view border condition. (**b**) Low-resolution mask in the intermediate image plane. (**c**) High-resolution mask in the intermediate image plane. The image was taken from: A. Borkowski, Z. Zalevsky and B. Javidi, "Geometrical Super Resolved Imaging Using Non periodic Spatial Masking," JOSA A 26, 589–601 (2009)

various physical parameters is tested. The sensitivity testing is performed for the random mask of the third kind.

The first test includes examining the sensitivity of the number of quantization bits in the camera. Figure 4.7a shows the reference images. Figure 4.7b illustrates the low-resolution images captured with a CMOS detector having a varied number of quantization bits. In Fig. 4.7c, the reconstructed image is depicted. One can see that it is not identical to the input, but due to the spatial masking, the obtained result is much closer to the original input not only by its spatial resolution but also by the gray level range obtained in each pixel.

From the figure, it is clear that even for a 4-bit camera, the image is fully reconstructed. From the preceding simulations, it is possible to conclude that the obtained result is not sensitive to the number of quantization bits of the CMOS detector. Even for a CMOS detector having two quantization bits, the original image is almost completely reconstructed.

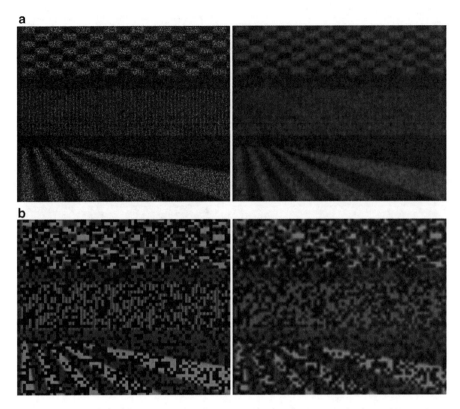

Fig. 4.5 Masked image (*left*) and blurred masked image (*right*) for: (**a**) high-resolution mask in the intermediate image plane and (**b**) low-resolution mask in the intermediate image plane. The image was taken from: A. Borkowski, Z. Zalevsky and B. Javidi, "Geometrical Super Resolved Imaging Using Non periodic Spatial Masking," JOSA A 26, 589–601 (2009)

The summary for the sensitivity of the described super resolution technique to the number of quantization bits is presented in Fig. 4.8. Figure 4.8a plots the standard deviation of the error (i.e., the difference) between the reconstructed and the original high-resolution images vs. the number of quantization bits. Obviously, reducing the number of bits increases the standard deviation of the error. One important parameter related to the presented approach is its numerical reliability which is related to the capability to perform the matrix inversion as indicated by (4.9). A way to test this reliability is to observe the condition number. This number is defined as follows:

$$\kappa = \text{NORM}(A^{\mathrm{T}}A + \alpha^2 I) \cdot \text{NORM}\left(\left(A^{\mathrm{T}}A + \alpha^2 I\right)^{-1}\right), \tag{4.10}$$

where NORM stands for the norm of a matrix (its largest singular value). In Fig. 4.8b, the condition number vs. number of quantization bits is presented.

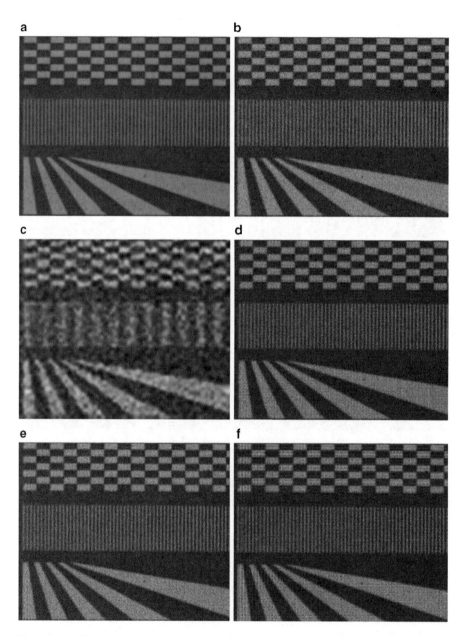

Fig. 4.6 (**a**) High-resolution reference. (**b**) High-resolution reference embedded with noise with a standard deviation of 20 gray levels. (**c**) The low-resolution image after blurring by a factor of 8 in every axis. (**d**) The reconstruction with mask, i.e., blocking the edges of the field of view. (**e**) The same as in (**d**) but with applying random binary mask with large pixels. (**f**) The same as in (**d**) but with applying random binary mask with small pixels. The size of the small pixel is 1/8 of the binned pixels of the CMOS detector (i.e., 11.5 μm). The size of the large pixel equals to the binned pixel of the CMOS detector (i.e., 92 μm). The image was taken from: A. Borkowski, Z. Zalevsky and B. Javidi, "Geometrical Super Resolved Imaging Using Non periodic Spatial Masking," JOSA A 26, 589–601 (2009)

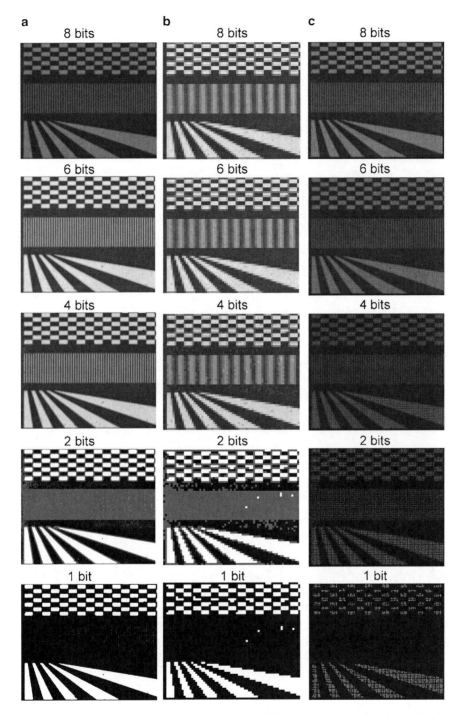

Fig. 4.7 Computer simulations that examine the sensitivity of the suggested technique to the number of quantization bits of the camera: (**a**) reference images captured by a CMOS detector with varied number of quantization bits, (**b**) low-resolution images, (**c**) the reconstructed images. The image was taken from: A. Borkowski, Z. Zalevsky and B. Javidi, "Geometrical Super Resolved Imaging Using Non periodic Spatial Masking," JOSA A 26, 589–601 (2009)

Fig. 4.8 Tolerance to quantization. (**a**) Standard deviation of the error between the reconstructed and the original high-resolution images vs. number of quantization bits. (**b**) The condition number vs. number of quantization bits. The image was taken from: A. Borkowski, Z. Zalevsky and B. Javidi, "Geometrical Super Resolved Imaging Using Non periodic Spatial Masking," JOSA A 26, 589–601 (2009)

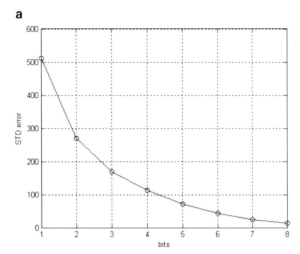

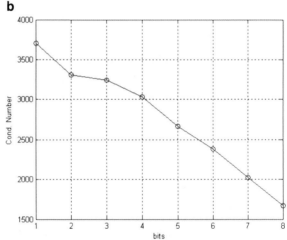

As anticipated, reduction in the number of bits increases the condition number which indicates lower reliability of the resulted reconstruction.

Figure 4.9 plots the qualitative effect of additive Gaussian, with zero mean, noise over the reconstruction quality. The noise level is described by its variance in a normalized dynamic range of 0–1. For instance, variance of 0.01 is equivalent to the standard deviation of 0.1 (square root of 0.01) which is the standard deviation of 25.5 ($=0.1 \times 255$) gray levels. The noise variance is between 0.0001 (very small noise level) and 0.1 (very strong noise). Figure 4.10 summarizes the obtained results. Figure 4.10a shows the standard deviation of the error (i.e., the difference) between the reconstructed and the original high-resolution images vs. noise. In Fig. 4.10b, condition number vs. noise level is plotted. Obviously, increasing the noise level increases the error of the reconstruction as well as the condition number (i.e., reducing the reliability of matrix inversion operation).

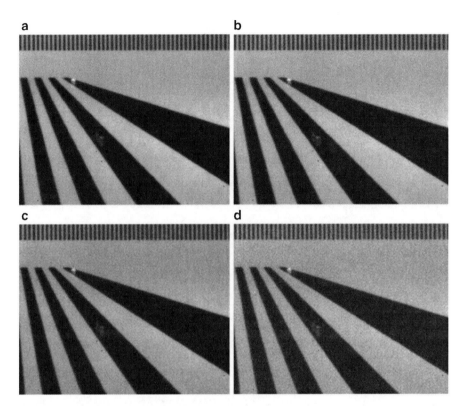

Fig. 4.9 Sensitivity to Gaussian noise (with zero average). The noise level is described by its variance in normalized dynamic range of 0–1. The noise variance is: (**a**) 0.0001, (**b**) 0.0002, (**c**) 0.0005, (**d**) 0.001, (**e**) 0.002, (**f**) 0.005, (**g**) 0.01, (**h**) 0.02, (**i**) 0.05, and (**j**) 0.1. The image was taken from: A. Borkowski, Z. Zalevsky and B. Javidi, "Geometrical Super Resolved Imaging Using Non periodic Spatial Masking," JOSA A 26, 589–601 (2009)

Next the effect of degree of transparency of the mask on the quality of reconstruction is evaluated. As the mask gets more transparent, less a priori known values could be inserted into the reconstruction algorithm. The images of Fig. 4.11 are obtained for random masks with different blocking portions. The range of the blocking portion used in test is between 0.0005 (0.05% of the image is blocked) and up to 0.5 (i.e., half of the image is blocked). Summary of the obtained results is seen in Fig. 4.12 – here the standard deviation of the error (i.e., the difference) between the reconstructed and the high-resolution blocked image images is shown.

One may see that for small blocking portions, the error is fixed and then when the blocking portion is increased the error is significantly reduced. Figure 4.12b presents the standard deviation of the error between the reconstructed and the original (unblocked) high-resolution images. Here, increase of the blocking increases the error since the reconstructed image becomes more different from the high-resolution original reference image. As previously described, this reduction in performance can easily be resolved by capturing two rather than a single

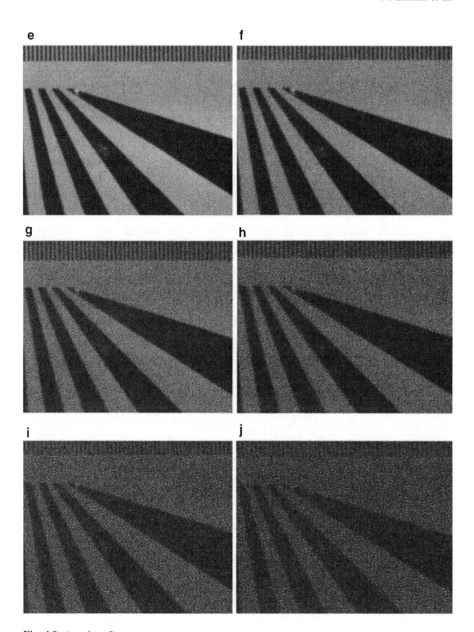

Fig. 4.9 (continued)

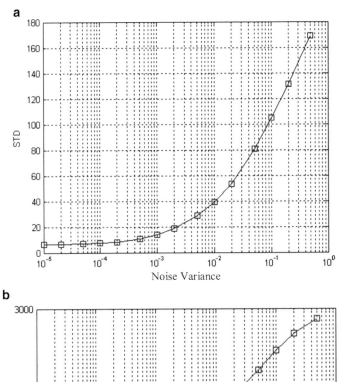

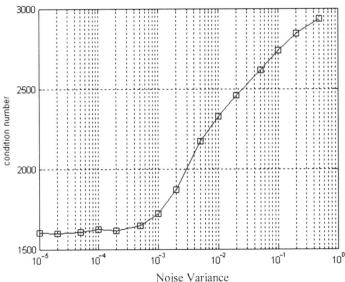

Fig. 4.10 Tolerance to noise. (**a**) Standard deviation of the error between the reconstructed and the original high-resolution images vs. noise level. (**b**) The condition number vs. noise level. The image was taken from: A. Borkowski, Z. Zalevsky and B. Javidi, "Geometrical Super Resolved Imaging Using Non periodic Spatial Masking," JOSA A 26, 589–601 (2009)

image while for each image a different random mask is used (while no spatial overlapping exists between the blocking locations in the two random masks). In Fig. 4.12c, the condition number vs. the blocked portion is examined. Increasing the blocking portion gives us more a priori knowledge and thus the condition number is reduced (meaning increased reliability of the inversion).

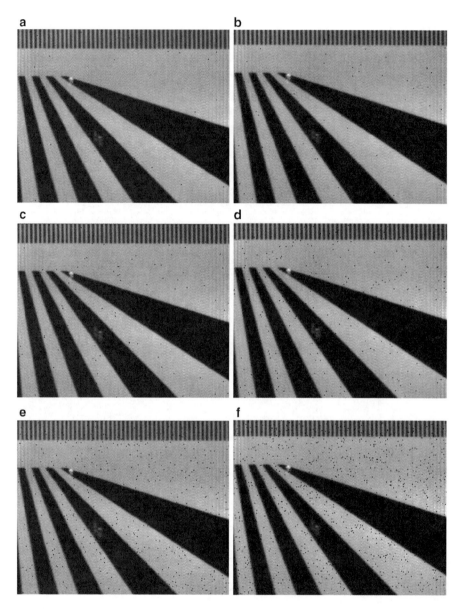

Fig. 4.11 Performance vs. the blocking portion of the random mask. The blocking portion is as follows: (**a**) 0.0005, (**b**) 0.001, (**c**) 0.002, (**d**) 0.005, (**e**) 0.01, (**f**) 0.02, (**g**) 0.05, (**h**) 0.1, (**i**) 0.2, and (**j**) 0.5. The image was taken from: A. Borkowski, Z. Zalevsky and B. Javidi, "Geometrical Super Resolved Imaging Using Non periodic Spatial Masking," JOSA A 26, 589–601 (2009)

g   h

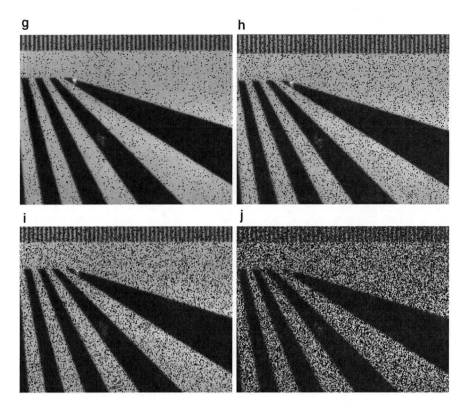

i   j

Fig. 4.11   (continued)

Figure 4.13 shows the results of the performance vs. the value of α coefficient in the Tikhonov's regularization algorithm [see (4.9)]. The range for the values of α coefficient is between 0.001 and up to 20.

Figure 4.14 summarizes the effect of the regularization coefficient – Fig. 4.14a presents the standard deviation of the error (i.e., the difference) between the reconstructed image and the high-resolution reference image. Increasing the coefficient contributes to increased reconstruction error. Figure 4.14b plots the condition number vs. the regularization coefficient. One may see that increasing the coefficient reduces the condition number (i.e., increases the reliability of the inversion).

4.2   Random Angular Coding for Super Resolved Imaging

4.2.1   Introduction

This section presents a super resolution work that generalizes the concept of super resolution based on two static gratings by using two random static masks for the encoding/decoding [11].

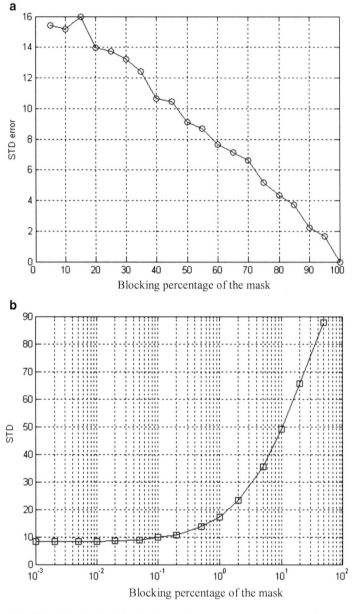

Fig. 4.12 The effect of partial blocking by the random mask over the overall performance. (**a**) The standard deviation of the error between the reconstructed and the high-resolution blocked image. (**b**) The standard deviation of the error between the reconstructed and the original (unblocked) high-resolution images. (**c**) The condition number vs. the blocked portion. The image was taken from: A. Borkowski, Z. Zalevsky and B. Javidi, "Geometrical Super Resolved Imaging Using Non periodic Spatial Masking," JOSA A 26, 589–601 (2009)

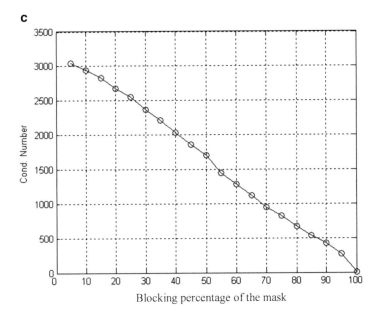

Fig. 4.12 (continued)

As a most noticeable fact, the super resolution effect is obtained without a cost neither in the time domain nor in the field of view domain. This is contrary to previous approaches which exploited time domain [12–20] and field of view [21–27] to achieve a higher resolving power. Now, the impact is only performed in dynamic range since the contrast of the obtained super resolved image is reduced. As in the case of static grating approaches, the random masks must have smaller features than those aimed to be resolved in the object. Here, the concept reported in ref. [28] is expanded to the two-dimensional (2-D) case while reporting the application of the method not only for coherent but also for incoherent (extended white light source) illumination. Moreover, the gain in resolution depends on the encoding mask pixel size and a factor of noise but it is independent on the NA of the imaging system. This is an important improvement in comparison to the original super resolving idea considering two fixed variable masks. The achieved experimental results suggest that the technique can be implemented in microscopy by properly selecting the pixel size of the encoding masks to the NA of the objective lens.

4.2.2   Mathematical Derivation

The schematic sketch of the proposed setup is shown in Fig. 4.15. For simplicity, a 1-D analysis is performed while the extension to 2-D is straightforward.

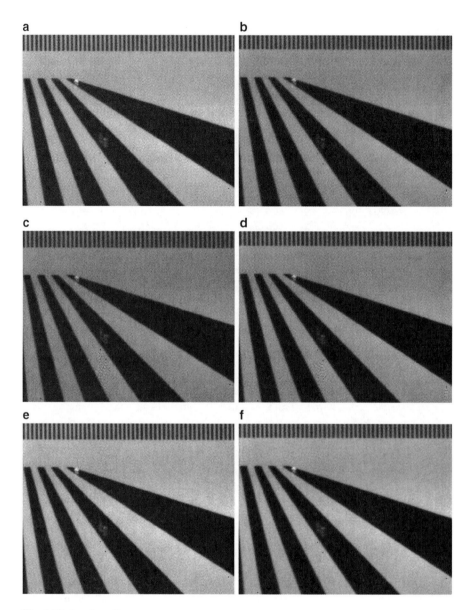

Fig. 4.13 (continued)

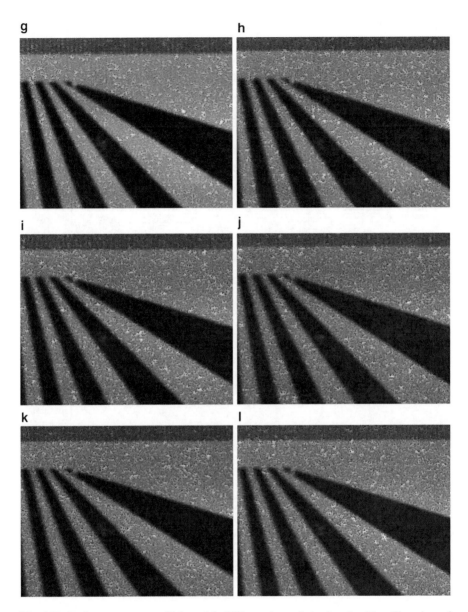

Fig. 4.13 Performance vs. α coefficient of the Tikhonov's regularization algorithm. The values of α are: (**a**) 0.001, (**b**) 0.002, (**c**) 0.005, (**d**) 0.01, (**e**) 0.02, (**f**) 0.05, (**g**) 0.5, (**h**) 1, (**i**) 2, (**j**) 5, (**k**) 10, and (**l**) 20. The image was taken from. A. Borkowski, Z. Zalevsky and B. Javidi, "Geometrical Super Resolved Imaging Using Non periodic Spatial Masking," JOSA A 26, 589–601 (2009)

Thus, the field distribution after free space propagation of z_1 equals to

$$g_z(x) = \int G(\mu) \exp\left(\pi i \lambda z_1 \mu^2\right) \exp(2\pi i x \mu) \mathrm{d}\mu, \qquad (4.11)$$

a

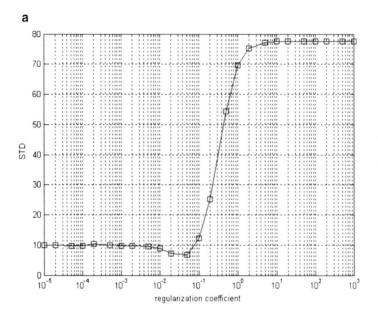

b

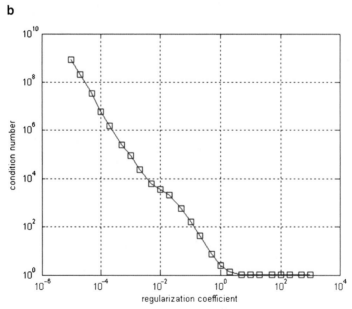

Fig. 4.14 The effect of the regularization coefficient over: (**a**) the standard deviation between the reconstructed image and the high-resolution reference image. (**b**) The condition number vs. the regularization coefficient. The image was taken from: A. Borkowski, Z. Zalevsky and B. Javidi, "Geometrical Super Resolved Imaging Using Non periodic Spatial Masking," JOSA A 26, 589–601 (2009)

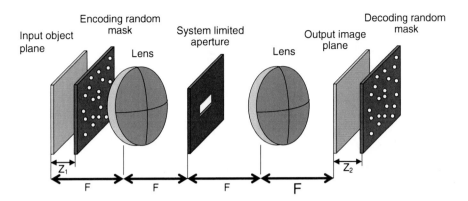

Fig. 4.15 Theoretical layout of the proposed setup. The image was taken from: D. Sylman, V. Micó, J. García and Z. Zalevsky, "Random Angular Coding for Superresolved Imaging," Appl. Opt. 49, 4874–4882 (2010)

where

$$G(\mu) = \int g(x) \exp(-2\pi i x \mu) dx. \tag{4.12}$$

This distribution is multiplied by the random encoding mask equals to $m(x)$ and the obtained product equals to

$$\int \left[\int M(\mu - \mu_1) G(\mu_1) \exp\left(\pi i \lambda z_1 \mu_1{}^2\right) d\mu_1 \right] \exp(2\pi i x \mu) d\mu, \tag{4.13}$$

while

$$M(\mu) = \int m(x) \exp(-2\pi i x \mu) dx. \tag{4.14}$$

Next back free space propagation of $-z_1$ is performed to obtain the field distribution in the input plane while the effect of the encoding mask is included

$$\int \left[\int M(\mu - \mu_1) G(\mu_1) \exp\left(\pi i \lambda z_1 \mu_1{}^2\right) d\mu_1 \right] \exp\left(-\pi i \lambda z_1 \mu^2\right) \exp(2\pi i x \mu) d\mu, \tag{4.15}$$

we switch now to the aperture plane by performing a Fourier transform

$$\int \left[\int \left(\int M(\mu - \mu_1) G(\mu_1) \exp\left(\pi i \lambda z_1 \mu_1{}^2\right) d\mu_1 \right) \exp\left(-\pi i \lambda z_1 \mu^2\right) \exp(2\pi i x \mu) d\mu \right]$$
$$\times \exp\left(-\frac{2\pi i \mu_2 x}{\lambda F}\right) dx, \tag{4.16}$$

which after mathematical simplification equals to

$$\left[\int M\left(\frac{\mu_2}{\lambda F} - \mu_1\right) G(\mu_1) \exp\left(\pi i \lambda z_1 \mu_1{}^2\right) d\mu_1 \right] \exp\left(-\pi i \lambda z_1 \left(\frac{\mu_2}{\lambda F}\right)^2\right). \qquad (4.17)$$

It is next multiplied by the aperture (assuming a rect function for the aperture)

$$\left[\int M\left(\frac{\mu_2}{\lambda F} - \mu_1\right) G(\mu_1) \exp\left(\pi i \lambda z_1 \mu_1{}^2\right) d\mu_1 \right] \exp\left(-\pi i \lambda z_1 \left(\frac{\mu_2}{\lambda F}\right)^2\right) \mathrm{rect}\left(\frac{\mu_2}{\Delta \mu_2}\right) \qquad (4.18)$$

and after additional optical Fourier transform, the distribution reaches the output plane (yet without taking into account the second decoding mask)

$$\int \left[\int M\left(\frac{\mu_2}{\lambda F} - \mu_1\right) G(\mu_1) \exp\left(\pi i \lambda z_1 \mu_1{}^2\right) d\mu_1 \right] \exp\left(-\pi i \lambda z_1 \left(\frac{\mu_2}{\lambda F}\right)^2\right) \mathrm{rect}\left(\frac{\mu_2}{\Delta \mu_2}\right)$$
$$\times \exp\left(\frac{-2\pi i \mu_2 x}{\lambda F}\right) d\mu_2.$$
$$\qquad (4.19)$$

Then, change of variables is performed such that $v = \mu_2/\lambda F$

$$\int \left[\int M(v - \mu_1) G(\mu_1) \exp\left(\pi i \lambda z_1 \mu_1{}^2\right) d\mu_1 \right] \exp\left(-\pi i \lambda z_1 v^2\right) \mathrm{rect}\left(\frac{v}{\Delta \mu_2/\lambda F}\right)$$
$$\times \exp(-2\pi i v x) dv.$$
$$\qquad (4.20)$$

Now it is need to add a free space propagation of z_2 in order to reach the random decoding mask. To do that the angular spectrum approach for computing the free space propagation is used, i.e., spectrum is multiplied by the chirp phase factor

$$\int \left[\int M(v - \mu_1) G(\mu_1) \exp\left(\pi i \lambda z_1 \mu_1{}^2\right) d\mu_1 \right] \exp\left(\pi i \lambda (z_2 - z_1) v^2\right) \mathrm{rect}\left(\frac{v}{\Delta \mu_2/\lambda F}\right)$$
$$\times \exp(-2\pi i v x) dv.$$
$$\qquad (4.21)$$

Now after propagating a free space distance of z_2, the distribution is multiplied by the decoding random mask $m^*(x)$. The Fourier of this mask equals to

$$m^*(x) = \left(\int M(\mu) \exp(2\pi i x \mu) d\mu \right)^* = \int M^*(-\mu) \exp(2\pi i x \mu) d\mu \qquad (4.22)$$

and the expression obtained equals to

$$\int \left[\int M(v - \mu_1) G(\mu_1) \exp\left(\pi i \lambda z_1 \mu_1{}^2\right) d\mu_1 \right] \exp\left(\pi i \lambda (z_2 - z_1) v^2\right) \mathrm{rect}\left(\frac{v}{\Delta \mu_2/\lambda F}\right)$$
$$\left[\int M^*(-\mu_2) \exp(2\pi i x \mu_2) d\mu_2 \right] \exp(-2\pi i v x) dv. \qquad (4.23)$$

It may be rewritten as a convolution in the Fourier domain

$$\int \int \left[\int M(v_1 - \mu_1)G(\mu_1)\exp\left(\pi i \lambda z_1 \mu_1{}^2\right)d\mu_1 \right] \exp\left(\pi i \lambda (z_2 - z_1)v_1{}^2\right)\mathrm{rect}\left(\frac{v_1}{\Delta\mu_2/\lambda F}\right)$$
$$M^*(-v + v_1)\exp(2\pi i x v)dv_1\,dv.$$

$$(4.24)$$

Now it is left to add additional free space propagation of $-z_2$, which means another Fourier transform multiplied by the chirp factor and inverse Fourier transform

$$\int \int \left[\int M(v_1 - \mu_1)G(\mu_1)\exp\left(\pi i \lambda z_1 \mu_1{}^2\right)d\mu_1 \right] \exp\left(\pi i \lambda (z_2 - z_1)v_1{}^2\right)\mathrm{rect}\left(\frac{v_1}{\Delta\mu_2/\lambda F}\right)$$
$$M^*(-v + v_1)\exp\left(-\pi i \lambda z_2 v^2\right)\exp(2\pi i x v)dv_1\,dv.$$

$$(4.25)$$

This is the field distribution in the output plane. Note that the masks of encoding and decoding are random and therefore are uncorrelated

$$\int M(v)M^*(v - v_1)dv = \delta(v_1).$$

$$(4.26)$$

This decorrelated relation is very strong (the mask is very random) and it may be rewritten as

$$\int f(v) \cdot M(v)M^*(v - v_1)dv = \delta(v_1)$$

$$(4.27)$$

for any general function $f(v)$.

Since the distributions are fields, M can be complex and nonhermitic. It is possible to rewrite (4.5) as

$$\int \int G(\mu_1)\exp\left(\pi i \lambda z_1 \mu_1{}^2\right)\exp\left(-\pi i \lambda z_2 v^2\right)\exp(2\pi i x v)\cdot$$
$$\left[\int \exp\left(\pi i \lambda (z_2 - z_1)v_1{}^2\right)\mathrm{rect}\left(\frac{v_1}{\Delta\mu_2/\lambda F}\right)M(v_1 - \mu_1)\,M^*(-v + v_1)dv_1 \right]d\mu_1\,dv$$

$$(4.28)$$

using the assumption of (4.27), yields

$$\int \int G(\mu_1)\exp\left(\pi i \lambda z_1 \mu_1{}^2\right)\exp\left(-\pi i \lambda z_2 v^2\right)\exp(2\pi i x v)\delta(v - \mu_1)d\mu_1\,dv$$

$$(4.29)$$

and results with

$$\int G(\mu_1)\exp\left(\pi i \lambda (z_1 - z_2)\mu_1{}^2\right)\exp(2\pi i x \mu_1)d\mu_1.$$

$$(4.30)$$

In the spatially coherent case, the expression for intensity is:

$$I(x) = \left| \int G(\mu_1) \exp\left(\pi i \lambda (z_1 - z_2) \mu_1{}^2\right) \exp(2\pi i x \mu_1) d\mu_1 \right|^2 \qquad (4.31)$$

For $z_1 = z_2$, one can obtain super resolution since the field of the output equals to the full resolution object's field $g(x)$. An interesting application for the proposed setup can be filtering. By choosing $z_1 - z_2$ not being equal to zero, a filtering operation is actually applied over the input object.

Note that for the assumption of (4.27), the Fourier transform of the encoding/decoding mask M must contain a lot of features which means that $m(x)$ should be large in the spatial domain, at least as large as $g(x)$ and definitely much larger than the PSF of the imaging system before super resolution (within the width of the aperture which is a rect in this case, the function M should have as much features as possible). In addition, the spectral width of the coding/decoding mask, i.e., the width of M should be as large as the synthetic aperture to be generated in the super resolution process. This in a way resembles CDMA coding where orthogonality is also required in order to separate mixed bits. In this case, the resolution of $m(x)$, i.e., its smallest feature, should at least as small as the smallest desired feature in $g(x)$ divided by the super resolution factor. This is the cost for the super resolution improvement in addition to the energy and contrast loses.

4.2.3   Numerical Simulation of the System

Three numerical simulations are presented in this subsection. One is for 1-D super resolution and two numerical simulations are for a 2-D resolving. For the system simulations, a spatially coherent illumination at a wavelength of 500 nm is assumed. The size of the pixels in the input mask is 0.1 mm, and the density of the random halls in the encoding/decoding mask is 25%. An example of an encoding/decoding mask that is used in the simulations is presented in Fig. 4.16.

In the 1-D simulation, the distances are chosen to be $z_1 = z_2 = 8$ m, the size of the low pass filter is 1.99 lines/mm. The width of the lines of the input object is 0.2 mm.

In the 2-D simulation, the distances are $z_1 = z_2 = 10$ m for the grating input and 12 m for the lattice input object. The size of the low pass filter is of 1.99 lines/mm in both dimensions. The width of the lines of the input grating is 0.28 mm and the size of the lattice unit is 0.2 mm $\times$ 0.2 mm.

Figure 4.17 presents the numerical simulations of the setup. In Fig. 4.17a, the image of the high-resolution reference is presented. Figure 4.17b shows the low-resolution reference as it is seen after the spatial blurring due to a low-resolution imaging system. After applying the proposed approach, the obtained result is seen in Fig. 4.17c. In Fig. 4.17d, one may see the results of Fig. 4.17c after reducing the additive noises added due to the processing procedure. One may see that the

Fig. 4.16 The mask that was used for the encoding and the decoding in the numerical simulation. The image was taken from: D. Sylman, V. Micó, J. García and Z. Zalevsky, "Random Angular Coding for Superresolved Imaging," Appl. Opt. 49, 4874–4882 (2010)

reconstructed image is very similar to the original high-resolution reference. The SNR of the image had been improved from 0.7 to 0.9.

Figure 4.18 shows two additional numerical simulations of the proposed technique for 2-D super resolution case. In Fig. 4.18a-I and a-II, two high-resolution input reference images is shown. In Fig. 4.18b-I and b-II, low-resolution references as they are seen after the spatial blurring due to a low-resolution imaging system is plotted. After applying the proposed approach by adding the random masks, the obtained results are seen in Fig. 4.18c-I and c-II. Figure 4.18d-I and d-II presents the obtained results of Fig. 4.18c-I and c-II after reducing the additive noise generated in the processing. One may see that the reconstructed images also in the 2-D case are very similar to the original high-resolution references, exactly as it is in the 1-D case. The SNR of the image had been improved from 0.66 in both images 4.18b-I and b-II to 1.3 at image 4.18d-I and 0.83 at image 4.18d-II.

4.2.4   Experimental Results

To validate the proposed approach working under incoherent illumination, the optical setup showed in Fig. 4.19 was assembled at the laboratory. Extended (nonpunctual) polychromatic (white light) illumination is provided by Fiber-Lite MI-150 fiber optic illuminator (halogen lamp source focused onto a fiber optic light guide). For the encoding/decoding process, two identical binary amplitude square random masks with different magnifications are used in the experiment. Figure 4.20

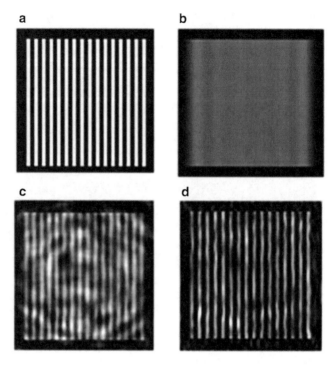

Fig. 4.17 Numerical results for 1-D super resolution. (**a**) The high-resolution reference image. (**b**) The image after reducing high spatial frequencies. (**c**) The recovered image. (**d**) The recovered image after reducing noise. The image was taken from: D. Sylman, V. Micó, J. García and Z. Zalevsky, "Random Angular Coding for Superresolved Imaging," Appl. Opt. 49, 4874–4882 (2010)

depicts the area of the masks which is used for encoding/decoding the object's angular information and where the black circle acts as a reference detail. The masks are fabricated using a standard process: (a) Taking samples made of glass substrate, 200 nm chrome layer and 500 nm photoresist layer. (b) Removing photoresist in desired locations using with photolithography methodology. (c) Removing chrome using chrome etching technique. The resulted mask is transparent in locations were chrome was etched. The encoding mask (M_1) has a pixel size of 3 μm and a total width of 4.5 mm while the decoding one (M_2) has a pixel size of 20 μm and a total width of 30 mm. Thus, the corresponding mask magnification is set to be 6.67.

Two imaging modules compose the experimental setup. In the first one, a variable circular diaphragm is attached to the back focal plane of a commercial microscope lens having 0.1 NA. The diaphragm allows us to stop down the resolution of the objective lens in order to match its NA with the size of M_1 used in the experiment. The magnification of the first imaging system must be properly adjusted to be equal to that one defined by both random masks. Otherwise, no super resolution effect will be attainable. To allow this, the first imaging system is placed onto a micrometer stage in order to allow magnification adjustment between the M_1 and M_2 planes.

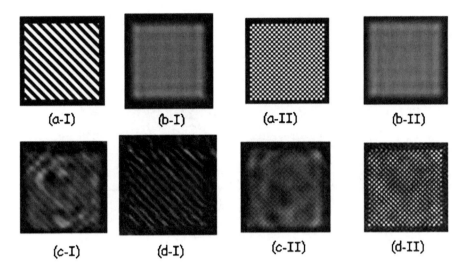

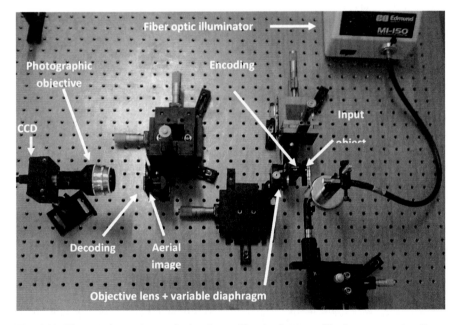

Fig. 4.18 Numerical results for 2-D super resolution: (I) 2-D input grating (II) 2-D input lattice. (**a**) The high-resolution reference image. (**b**) The image after reducing high spatial frequencies. (**c**) The recovered image. (**d**) The recovered image after reducing noise. The image was taken from: D. Sylman, V. Micó, J. García and Z. Zalevsky, "Random Angular Coding for Superresolved Imaging," Appl. Opt. 49, 4874–4882 (2010)

Fig. 4.19 The experimental setup for incoherent illumination case. The image was taken from: D. Sylman, V. Micó, J. García and Z. Zalevsky, "Random Angular Coding for Superresolved Imaging," Appl. Opt. 49, 4874–4882 (2010)

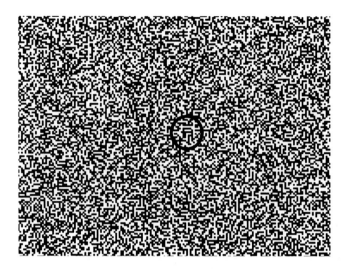

Fig. 4.20 Picture of the theoretical design of the random mask used in the experiment (only a small part is included). The *black circle* marks a mask's detail that can also be traced in Fig. 4.21 for reference. The image was taken from: D. Sylman, V. Micó, J. García and Z. Zalevsky, "Random Angular Coding for Superresolved Imaging," Appl. Opt. 49, 4874–4882 (2010)

Figure 4.21 depicts the cases without and with proper magnification matching between the masks. The white circle is for referencing both images and also for Fig. 4.20. Since the input object is placed before M_1, its image will be placed also in a plane previous to plane M_2. Thus, the second imaging module images the aerial image provided by the first system through M_2. A photographic objective with variable focus (or magnification) is selected as second imaging module to magnify the aerial image into the output plane where the CCD (Basler A312f, 582×782 pixels, 8.3 μm pixel size, 12 bits/pixel) is placed. This second imaging module plays the role of the tube lens used in microscope systems. Due to the magnification ratio between the two imaging modules of the setup, M_2 could be a low-frequency mask (higher pixel size than M_1, as it was previously described) and no need for high-resolution optics which is necessary in the second imaging module.

Under these assumptions, the super resolution approach is applied. A positive USAF resolution test target is used as input object. The circular diaphragm of the first imaging lens is closed in order to stop down the resolution of the experimental setup. Figure 4.22a depicts the low-resolution image provided by the experimental setup where Group 6 – Element 3 (G6-E3 from now on) is the last resolved element in the test that defines a resolution limit which equals to 12.4 μm (80.6 lp/mm). This resolution limit corresponds with a theoretical value of 0.022 NA in the first imaging module considering the central wavelength (0.55 μm) of the illumination. After performing the super resolved approach, the resolution is improved until G7-E2 corresponding with 6.9 μm (144.0 lp/mm) as shown in Fig. 4.22b, which defines that a resolution gain factor is equal to 1.8.

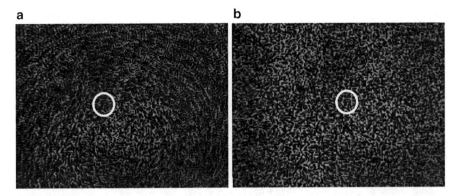

Fig. 4.21 Example of: (**a**) magnification mismatch and (**b**) perfect magnification adjustment between the two masks. *White circle* marks the same area for reference. The image was taken from: D. Sylman, V. Micó, J. García and Z. Zalevsky, "Random Angular Coding for Superresolved Imaging," Appl. Opt. 49, 4874–4882 (2010)

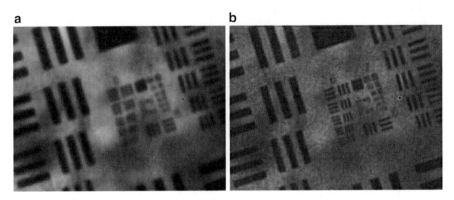

Fig. 4.22 Experimental results: (**a**) without and (**b**) with using the proposed approach and corresponding with conventional low-resolution and superresolved images, respectively. The image was taken from: D. Sylman, V. Micó, J. García and Z. Zalevsky, "Random Angular Coding for Superresolved Imaging," Appl. Opt. 49, 4874–4882 (2010)

Since the pixel size in M_1 has a width of 3 μm, the expected theoretical resolution limit is twice the pixel width, i.e., 6 μm (or 166.7 lp/mm). This resolution limit corresponds in the USAF test with G7-E3 (161 lp/mm) which is very close to the theoretical limit and it is not resolved due to experimental factors such as noise, contrast reduction, mismatch between masks, etc. But in any case, this is the best resolution limit that can be achieved using the proposed approach: that one can define by the minimum period of the encoding mask. And such minimum resolution limit is theoretically independent on the NA of the first imaging module. Then, the purpose is to demonstrate this theoretical assumption. Figure 4.23 depicts the experimental results achieved with different diameters of the limiting diaphragm. Running from left to right, the NA value is increased from 0.016 to 0.022 and to 0.031, and the

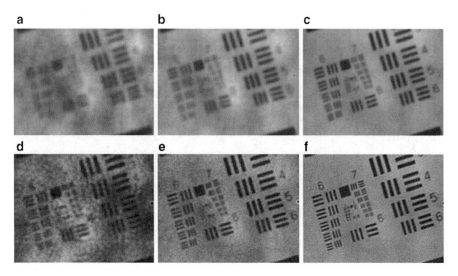

Fig. 4.23 Experimental results showing that the gain in resolution does not depend on the NA of the first module imaging system. (**a**)–(**d**), (**b**)–(**e**), and (**c**)–(**f**) depict different cases of low and superresolved images corresponding with different diameters of the lens diaphragm. The image was taken from: D. Sylman, V. Micó, J. García and Z. Zalevsky, "Random Angular Coding for Superresolved Imaging," Appl. Opt. 49, 4874–4882 (2010)

resolution limit is improved from 17.5 µm (G5-E6 in Fig. 4.23a), 12.4 µm (G6-E3 in Fig. 4.23b), and 8.8 µm (G6-E6 in Fig. 4.23c) to 6.9 µm (G7-E2 in Fig. 4.23d) and 6.2 µm (G7-E3 in Fig. 4.23e–f). And the corresponding resolution gain factors are 2.5, 2, and 1.4, respectively. Thus, it is demonstrated that the resolution limit of the setup is defined by the minimum pixel size of the encoding mask M_1.

References

1. Borkowski, A., Zalevsky, Z., Javidi, B.: Geometrical super resolved imaging using non periodic spatial masking. J. Opt. Soc. Am. A **26**, 589–601 (2009)
2. Fortin, J., Chevrette, P., Plante, R.: Evaluation of the microscanning process. SPIE Vol. 2269, Infrared Technology XX, Bjorn F. Andresen, Editors, 271–279 (1994)
3. Borman, S., Stevenson, R.: Super-resolution from image sequences – A review. In: Proceedings of the 1998 Midwest Symposium on Circuits and Systems, Notre Dame, IN, USA, pp. 374–378 (1998)
4. Borman, S.: Topics in multiframe superresolution restoration. Ph.D. dissertation, University of Notre Dame (2004)
5. Stern, A., Javidi, B.: Random projections imaging with extended space-bandwidth product. IEEE/OSA J Disp Technol **3**(3), 315–320 (2007)
6. Bagheri, S., Javidi, B.: Extension of depth of field using amplitude and phase modulation of the pupil function. Opt. Lett. **33**(7), 757–759 (2008)

7. Martínez-Cuenca, R., Saavedra, G., Martínez-Corral, M., Javidi, B.: Extended depth-of-field 3-D display and visualization by combination of amplitude-modulated microlenses and deconvolution tools. IEEE J. Disp. Technol. **1**(2), 321–327 (2005)
8. Martinez, L., Javidi, B.: Synthetic aperture single-exposure on-axis digital holography. Opt. Express **16**(1), 161–169 (2008)
9. Hong, M., Kang, M.G., Katsaggelos, A.K.: An iterative weighted regularized algorithm for improving the resolution of video sequences. IEEE Int. Conf. Image Process. **2**, 474–477 (1997)
10. Vogel, C.R.: Non-convergence of the L-curve regularization parameter selection method. Inverse Probl **12**, 535–547 (1996)
11. Sylman, D., Micó, V., García, J., Zalevsky, Z.: Random angular coding for superresolved imaging. Appl. Opt. **49**, 4874–4882 (2010)
12. Françon, M.: Amélioration the résolution d'optique. Il Nuovo Cimento **Suppl. 9**, 283–290 (1952)
13. Lukosz, W.: Optical systems with resolving powers exceeding the classical limits. J. Opt. Soc. Am. **56**, 1463–1472 (1967)
14. Shemer, A., Mendlovic, D., Zalevsky, Z., García, J., García-Martínez, P.: Superresolving optical system with time multiplexing and computer decoding. Appl. Opt. **38**, 7245–7251 (1999)
15. Mendlovic, D., Lohmann, A.W., Konforti, N., Kiryuschev, I., Zalevsky, Z.: One dimensional super resolution optical system for temporally restricted objects. Appl. Opt. **36**, 2353–2359 (1997)
16. Mendlovic, D., Kiryuschev, I., Zalevsky, Z., Lohmann, A.W., Farkas, D.: Two dimensional super resolution optical system for temporally restricted objects. Appl. Opt. **36**, 6687–6691 (1997)
17. Mendlovic, D., Farkas, D., Zalevsky, Z., Lohmann, A.W.: High-frequency enhancement by an optical system for superresolution of temporally restricted objects. Opt. Lett. **23**, 801–803 (1998)
18. Shemer, A., Zalevsky, Z., Mendlovic, D., Konforti, N., Marom, E.: Time multiplexing super resolution based on interference grating projection. Appl. Opt. **41**, 7397–7404 (2002)
19. Zalevsky, Z., García, J., Micó, V.: Transversal superresolution with noncontact axial movement of periodic structures. J. Opt. Soc. Am. A **24**, 3220–3225 (2007)
20. Mico, V., Limon, O., Zalevsky, Z., García, J.: Transversal resolution improvement using rotating-grating time-multiplexing approach. J. Opt. Soc. Am. A **25**, 1115–1129 (2008)
21. Grimm, M.A., Lohmann, A.W.: Super resolution image for 1-D objects. J. Opt. Soc. Am. A **56**, 1151–1156 (1966)
22. Lukosz, W.: Optical systems with resolving powers exceeding the classical limits. II. J. Opt. Soc. Am. **57**, 932–941 (1967)
23. Bachl, A., Lukosz, W.: Experiments on superresolution imaging of a reduced object field. J. Opt. Soc. Am. **57**, 163–164 (1967)
24. Bartelt, H., Lohmann, A.W.: Optical processing of one-dimensional signals. Opt. Commun. **42**, 87–91 (1982)
25. Zalevsky, Z., Mendlovic, D., Lohmann, A.W.: Super resolution optical systems for objects with finite sizes. Opt. Commun. **163**, 79–85 (1999)
26. Sabo, E., Zalevsky, Z., Mendlovic, D., Komforti, N., Kiryushev, I.: Superresolution optical system with two fixed generalized Damman gratings. Appl. Opt. **39**, 5318–5325 (2000)
27. Sabo, E., Zalevsky, Z., Mendlovic, D., Komforti, N., Kiryushev, I.: Superresolution optical system using three fixed generalized gratings: experimental results. J. Opt. Soc. Am. A **18**, 514–520 (2001)
28. Zalevsky, Z., Eckhouse, V., Konforti, N., Shemer, A., Mendlovic, D., García, J.: Super resolving optical system based on spectral dilation. Opt. Commun. **241**, 43–50 (2004)